THE COLLEGE OF CHINESE MEDICINE

Practical Applications of Chinese Acupuncture, Herbal Medicine and Diagnosis

PAUL BRECHER BA FAcS MPCHM
PRINCIPAL OF THE COLLEGE OF CHINESE MEDICINE

and

PAUL ROBIN
HEAD OF TCM AT THE COLLEGE OF CHINESE MEDICINE
AND CHAIRMAN OF THE ACUPUNCTURE SOCIETY.

www.ccmlondon.com

FIRST EDITION 2010 BY CCM PRESS

SECOND EDITION 2011 BY CCM PRESS

Disclaimer

All training should be under the supervision of a qualified instructor of the College of Chinese Medicine. All techniques, theories, formulae and training methods, and any other information contained in any part of this book are presented for information purposes only! No liability whatsoever will be accepted, by the authors or publisher for any damage or injuries that may arise from the use or misuse of information contained in this book.

Acknowledgements

We would like to thank James Gardner for his help in the preparation of this book and his time and effort in editing and organising the text. We would also like to thank Phil Dodshon for proof reading the text and for his illustrations.

Preface

This book is for students of The College of Chinese Medicine to complement and support their lectures and clinical studies.

When first exposed to Traditional Chinese Medicine (TCM) it seems as if there is more information than one can ever possibly learn. There are over 500 herbs and over 400 acupuncture points in TCM however for everyday clinical practice we are able to treat most conditions with about 200 herbs and 200 acupuncture points.

So for those students who are prepared to study every day and dedicate themselves to the subject it is possible to make steady progress over time and eventually to become a confident and competent practitioner.

This book contains much of the core information that students of CCM need to know, please read it and contemplate it, add your own notes to it and highlight important points. All the information in this book is relevant to clinical practice.

Good luck in your studies

Paul Brecher

Paul Brecher BA FAcS MPCHM
Principal of The College of Chinese Medicine

Contents

Nomenclature

Abbr.	Explanation
Δ	Use Moxa Cones
BL	Bladder Meridian
CV	Conception (Ren) Meridian
EPA	Exogenous Pathogenic Attack
GB	Gall Bladder Meridian
GI	Gastrointestinal Tract
GU	Genital Urinary System
GV	Governing (Du) Meridian
H	Heart Meridian
HBP	High Blood Pressure
HT	Hua Tuo Paravertebral Points
K	Kidney Meridian
LBC	Lower Body Cavity
LI	Large Intestine Meridian
LIV	Liver Meridian
LU	Lung Meridian
MBC	Middle Body Cavity
P	Pericardium Meridian
QP	Quick Puncture
SB	Squeeze Blood
SI	Small Intestine Meridian
SP	Spleen Meridian
ST	Stomach Meridian
TCM	Traditional Chinese Medicine
TW	Triple Warmer (San Jiao) Meridian
UBC	Upper Body Cavity

Explanation of Nomenclature

EPA is an abbreviation for endogenous/exogenous pathogenic attack of Wind, Heat, Cold, Damp and Dryness which are elemental or environmental pathogens. EPA also refers to viruses and bacteria which are physical pathogens which need to be expelled from the body, for this we use Da Huang. This herb is a purgative to clear the viral and bacterial pathogens from the system. It is also used to clear any toxins from the system that could cause skin conditions like urticaria, psoriasis and eczema.

If we use the acupuncture pen to bleed an acupuncture point then SB will be written after the acupuncture point name, meaning Squeeze Blood. If we need to do a quick punch then QP will be used after the point name. To indicate that we should press moxa sticks on the handle of the needle, the symbol '/' will be put after the relevant acupuncture point. To indicate that we should put moxa cones on the acupuncture point, the symbol Δ will be put after the relevant acupuncture point. Also if three moxa cones need to be used then it will be shown like this 3 Δ. If a point needs to have a suction cup applied to it then after the point name will be written the word 'Cup'.

Chapter 1

Energy – *Qi*

1.1 A Chinese explanation of *qi*

There are many different types of *qi* in the body which transform into one another:

- **Inherited *qi*** is the *qi* we inherit from our parents and our ancestors. It is associated with the Kidneys, the lumbar vertebrae and the lower Dan Tian energy centre located just below the navel at CV3 and CV4.
- **Acquired *qi*** is derived from a combination of Nutrient *qi* and Clear *qi*.
- **Nutrient *qi*** comes from the food we eat.
- **Clear *qi*** comes from the air we breath.

In some ancient documents Inherited *qi* is referred to as Pre-Birth *qi* and Acquired *qi* is referred to as Post-Birth *qi*.

Before we are born we rely on our Inherited *qi* and after we are born we need Acquired *qi*. The Acquired *qi* combines with a little of the Inherited *qi* on each breath we take and makes the Meridian *qi* which keeps the body alive.

When the Inherited *qi* has run out we die. Even if we try and get more Acquired *qi* it will not help us.

The Spleen extracts Nutrient *qi* from the food in the Stomach. It sends this upward to nourish the Heart and Lungs. The Spleen also uses it to make blood to nourish the whole body.

The Nutrient *qi* combines in the chest with the Clear *qi* inhaled from the air to make Pectoral *qi*. It is stored in the chest and promotes the functions of the Lung in controlling respiration, as well as promoting the function of the Heart in controlling the circulation of blood.

All the different types of *qi* in the body are called Constructive *qi* and this supports the Defensive *qi* which is on the surface of the body and protects it from EPA, Exogenous Pathogenic Attack.

If the Constructive *qi* is weak it will fail to support the Defensive *qi* and so we will be vulnerable to EPA and become ill.

So we need to eat the right amount of the appropriate food at the right times, get enough sleep at the right time and do a bit of exercise every day then we will be healthy and strong.

1.2 A western explanation of *qi*

Lecture given by Paul Brecher BA FAcS MPCHM The Principal of the College of Chinese Medicine at the NHS Conference on Alternative and Complementary Medicine in Primary Care in London 29 October 2007

> Today I would like to talk to you about the possible mechanisms through which acupuncture works to heal the body. First I would like to put acupuncture in an historical perspective. In 1968 in China an ancient tomb from the Han Dynasty dating from 113 B.C. was excavated. Among the relics discovered were four acupuncture needles made of gold and five of silver. So acupuncture has been in continual use for over two thousand years, however this great age does not prove acupunctures effectiveness. It was not until 1979 that the World Health Organisation formally announced that acupuncture can be used to treat over forty different diseases.
>
> The World Health Organisation has since then published the following books:
>
> 1993 Standard International Acupuncture Nomenclature
> 1995 Guidelines for clinical research in Acupuncture
> 1999 Guidelines on Basic Training and Safety in Acupuncture
> 2002 Analysis of Reports on Controlled Clinical Trials of Acupuncture
>
> I would just like to read to you a small paragraph from this book 1995 WHO Guidelines for clinical research in acupuncture.
>
> > *The guidelines aim to encourage the use of systemic laboratory and clinical studies as a way of validating acupuncture, improving its acceptability to modern medicine, and thus extending its use as a simple, inexpensive, and effective therapeutic option. It sets out guidelines that incorporate the established methods and procedures of scientific investigation. The guidelines respond to both growing interest in the therapeutic applications of acupuncture and the need to validate these applications through the compilation of reliable and comparable clinical data.*
>
> With this in mind I would today like to discuss both the Western and Chinese explanations for how acupuncture works to heal the body. Both directly on the site of an injury or infection and also how acupuncture can indirectly have a controlling or healing effect on a distant part of the body some distance away from the area that is being needled.
>
> So first a possible western explanation of how acupuncture works locally on the site of an area of inflammation or infection.
>
> It would seem that acupuncture can increase microcirculation due to vasodilatation resulting in increased activity of the macrophages and leucocytes which through the process of phagocytosis can destroy bacteria and other harmful toxins and inflammatory components.

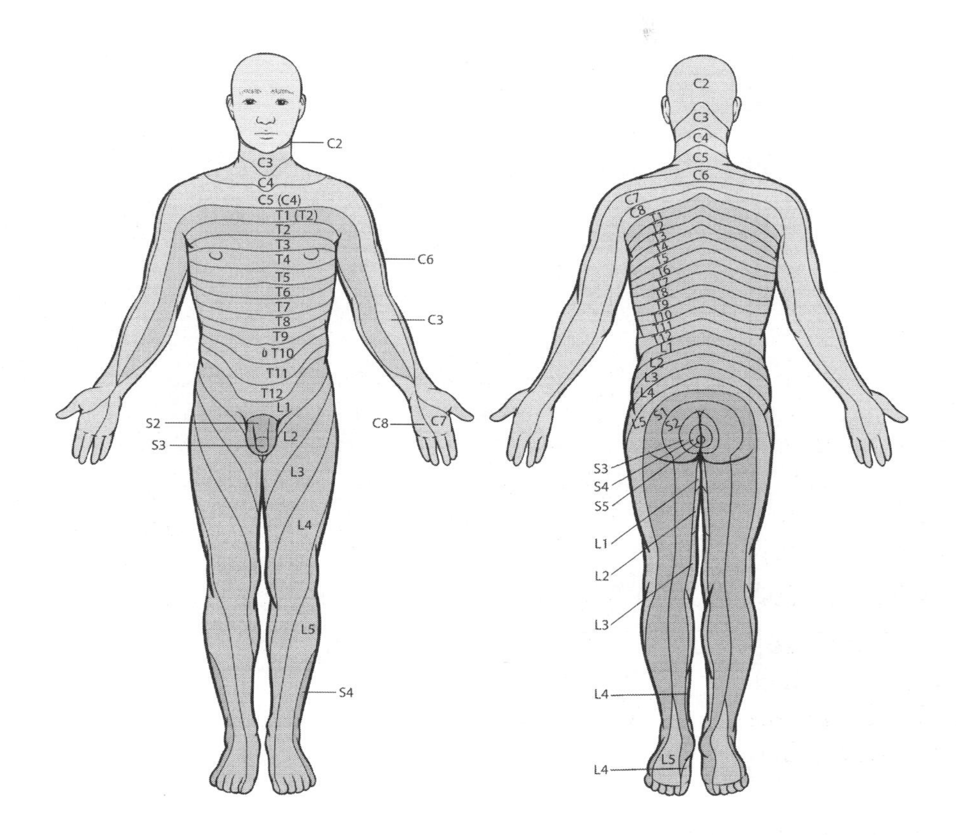

Figure 1.1: Diagram of Dermatomes

And now a possible western explanation of how acupuncture can effect areas of the body that are distant from where the needles are inserted. The mechanism could be via the dermatome zones which are the areas of sensory innervations on the skin of each spinal nerve.

It is possible that through this mechanism the surface stimulus created by acupuncture may effect distant areas of the body and in addition through the deep layer of the nerve may also influence organ function as well.

These mechanisms that I have described so far may also initiate a cascade of further processes in the body that further increase the bodies own self healing ability.

For example acupuncture may also induce an increase in the anabolic phase of the metabolism through influencing the activity of the parasympathetic nervous system. This would promote tissue regeneration, counter immune deficiency and reduce inflammation and pain.

There is also a possibility that acupuncture may affect the nerves in such a way as to influence the brain stem at the top of the spinal cord and regulate the hypothalamus which is just above it.

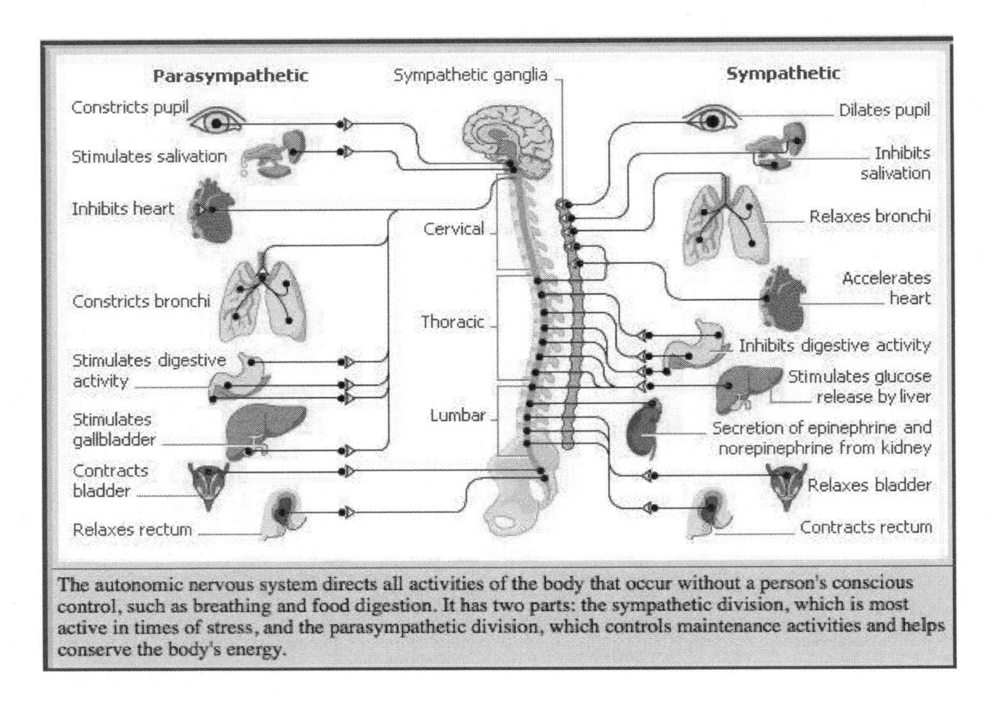

Figure 1.2: Diagram of the Sympathetic and Parasympathetic Nervous System

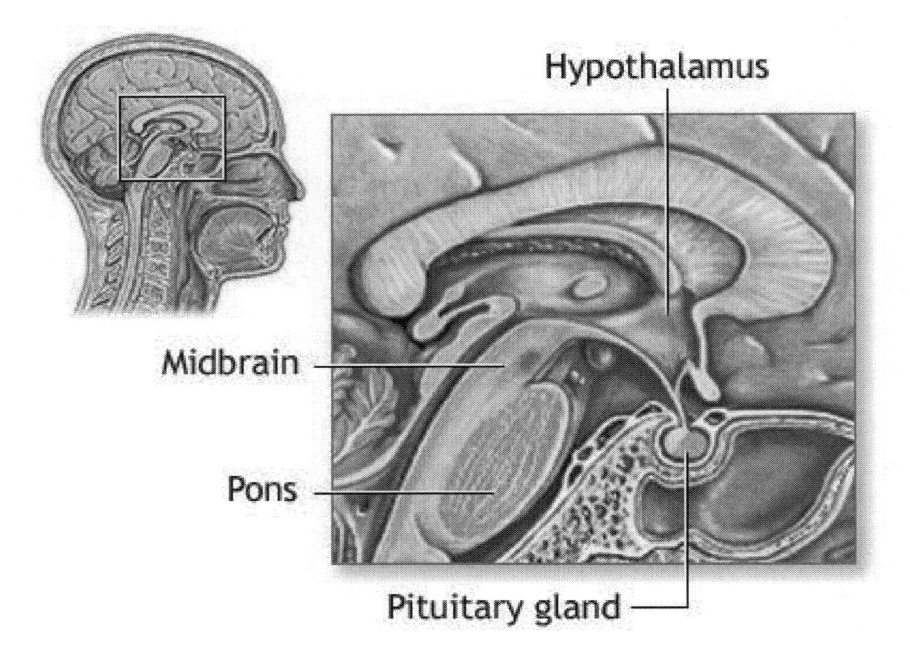

The hypothalamus links the nervous system to the endocrine system via the pituitary gland and is responsible for producing endorphin which has an analgesic effect.

As far as I know these are the current ideas which need to be further researched for us to understand the mechanisms of acupuncture from a western perspective.

I would now like to discuss acupuncture from the Chinese perspective. Measurements of the acupuncture points and acupuncture meridian pathways have found that they are the areas of least resistance to electrical conductivity.

There are a number of process by which these electrical signals travels through the body along the meridians. When it travels through the nervous system it is through the activity of the neurons. Through the body fluids by the ionic transfer of electrons. And it travels through the tissue via the cell membrane which has got selective protein pumps which can pump positive ions to one side and negative ions to the other, creating a potential difference across the membrane, creating a current.

In both China and the West patients have had brain scans whilst undergoing acupuncture treatment of the feet and specific areas of the brain have shown increased activity responding to acupuncture on different meridian acupuncture points.

So I would theorise that one of the mechanisms of acupuncture is to send electrical signals via the meridian pathways to the brain which then increase or reduce certain processes in the body and regulate organ function. More research is needed to clarify the various possible mechanisms that I have outlined here today.

Chapter 2

Traditional Chinese Medical Anatomy

2.1 The Internal Organs

2.1.1 The Six Full Organs

The Lungs, Spleen, Heart, Kidneys, Pericardium and Liver are the Full Organs which make and store substances. The Lungs make and store Qi, the Kidneys make and store Essence, the Spleen makes Blood, the Liver filters and stores Blood and the Heart pumps the Blood. And the Pericardium stores the Heart.

2.1.2 The Six Hollow Organs

The Large Intestine, Stomach, Small Intestine, Bladder, Triple Warmer and Gall Bladder are Hollow Organs which receive, digest and absorb food and transmit and excrete waste. The Stomach receives food, the Small and Large Intestine transmit and excrete waste. The Bladder releases urine and the Triple Warmer releases heat, and the Gall Bladder releases bile.

We should make sure the full organs are full and the hollow organs are emptying without obstruction and that the *qi* is flowing through the whole meridian system smoothly.

2.1.3 Extraordinary Organs

The Brain and Blood Vessels are connected to the Heart. The Marrow, Brain and Bones and the Uterus (Womb) are connected to the Kidneys.

2.2 The Three Body Cavities

When there is a problem in a body cavity it can affect all the organs in that cavity, so when we are thinking about treatment we are not only considering which organs to treat but also how to treat the whole of the particular body cavity that has been affected.

2.2.1 The Upper Body Cavity

The Upper Body Cavity UBC is the area above the solar plexus, it contains the Lungs, Heart, Pericardium and the Brain.

2.2.2 The Middle Body Cavity

The Middle Body Cavity MBC is the area between the solar plexus and the navel, it contains the Spleen, Stomach, Liver and Gall Bladder.

2.2.3 The Lower Body Cavity

The Lower Body Cavity LBC is the area below the navel, it contains the Large and Small Intestine, the Kidneys, Uterus (the Womb) and Bladder.

2.3 The Acupuncture Points

In the clinical practice of acupuncture we use the points on the twelve main organ meridians and points on the Governing and Conception Meridians. We also use the 'extra points', some of which are on meridians and some of which are not.

Acupuncture needles can be inserted on the location of a problem whether there is a point there or not as long as the direction of *qi* flow in the nearest meridian is taken into consideration. Against the meridian to reduce an excessive condition and in the direction to strengthen a deficiency.

The End and Beginning Points of the Meridians

For cases of acute conditions relating to that meridian or organ we use the end and beginning points of the meridians. For example in the case of sudden onset of an exogenous pathogenic attack (EPA) on the Lungs with heat and sore throat we squeeze blood (SB) on LU.11 and LI.1.

Additionally for pain, swelling, inflammation or arthritis on the hands or feet the end of the meridian point that went through the affected area would also be used.

2.3.1 *yin-yang* Paired Channels

The meridians connect the organs together in pairs. In addition, there are other ways in which the paired organs are connected;

Lung and Large Intestine Meridians

When a person has a severe cough in their Lungs, their anus (the end of the Large Intestine) spasms. To stop a dry cough we reduce LI.4, LU.10 and punch and squeeze blood from LU.11 and LI.1.

Stomach and Spleen Meridians

The nutrients from the food in the Stomach go to the Spleen were they are used to make the blood. To strengthen the Stomach we reinforce ST.36 (a commonly used herb to strengthen the Stomach and digestion is Bai Zhu). To clear food stagnation in the Stomach we reduce ST.36 (a commonly used herb to clear stagnation is Shen Qu). To strengthen a weak Spleen we reinforce SP.6. (a commonly used herb to strengthen the Spleen is Dang Shen (Codonopsis).

Heart and Small Intestine Meridians

Indigestion causes heartburn. This is heat rising up from the Small Intestine to the Heart (a commonly used herb to create a downwards movement in the chest and digestive system is Ban Xia. Reducing CV.12 and CV.13 would also counteract indigestion and heartburn).

Bladder and Kidney Meridians

Sudden fear weakens the Kidneys which weakens the Bladder causing involuntary urination. The main torso control point for the Kidneys is BL.23, we reinforce this point to strengthen the Kidneys. The main distant point to strengthen the Kidneys is KID.3. (a commonly used herb to astringe the LBC to counter involuntary urination is Fu Pen Zi).

Pericardium and Triple Warmer Meridians

Heat (fire) rises up through the body and so affects the Heart and Pericardium. So we clear heat from the whole body with TW.5 reducing and this will clear heat from and calm the Heart/Pericardium. (a commonly used herb to clear heat from the whole body is Huang Qin) We also directly clear heat from the Heart/Pericardium with H.7 reducing and/or P.6, in emergencies like extreme UBC heat damaging the brain we could punch H.8 and P.8 in the reducing direction as well and punch and squeeze blood (SB) from H.9 and P.9. (a commonly used herb to clear heat from the Heart is Zhu Ru)

Gall Bladder and Liver Meridians

The Liver secretes bile into the Gall Bladder which then secretes it into the intestines to help digest food. If there is heat in the Liver we can clear it by clearing heat from the Gall Bladder by reducing GB.34. We also directly clear heat from the Liver by reducing LIV.3. (a commonly used herb to clear heat from the Liver and Gall Bladder is Zhi Zi)

2.3.2 Other connections between the Organs

Lungs and Heart

The Heart and Lungs are both in the Upper Body Cavity, the oxygen inhaled into the Lungs goes to the Heart which then pumps oxygen enriched blood around the body. To strengthen the Lungs and Heart we can reinforce LU.1 and KID.27. If an exogenous pathogenic factor enters the Lungs it often then affects the Heart. To clear EPA from the Lungs we use LU.7 reducing. To calm the Heart we use H.7 reducing.

Lungs and Kidneys

The Lungs control exhalation and the Kidneys control inhalation. If a person has difficulty inhaling we strengthen the kidney *yang* - functional power, this will strengthen the *yang* - functional power of the Lungs. The point we use to strengthen the Kidneys to strengthen the Lungs is KID.6 reinforcing.

The Liver, Spleen and Heart

The Spleen makes the blood, the Liver stores and filters the blood and the Heart pumps the blood. When a patient has a blood problem we not only treat the blood but also look at the Spleen, Liver and Heart. deficiency of blood in one of these organs can create deficiency in another. To strengthen blood we reinforce the Spleen Shu (control point) on the back BL.20 and reinforce the distant point SP.6. The Liver and kidney meridians also pass through SP.6 so they are also being reinforced.

The Spleen and the Lungs

The nutrient *qi* that the Spleen acquires from the Stomach is sent up to strengthen the Lungs. This upwards movement can be increased by reinforcing SP.6 combined with reducing LI.4.

Heart (Fire) and Kidneys (Water)

Traditionally in TCM it is said that the Heart -fire and the kidney-water balance each other. In western medicine hormone secretions (essence) from the adrenal cortex of the Kidneys control the nerves that control the Heart. If a person has an erratic weak Heart we not only strengthen the Heart but also the Kidneys as well so that they can help regulate the Heart. A common treatment to balance the Heart and Kidneys for both physical and stress/emotional conditions is Hua Tuos points below the fifth thoracic vertebrae T5 reducing and BL.23 (this is the back Shu point of the Kidneys) reinforcing (balanced with GV.4 reducing)

Liver and Kidneys

The Liver is the storehouse of blood and the kidney is the storehouse of essence. Excess blood in the Liver is converted into essence in the Kidneys. Excess essence is converted into blood in the Liver. Great loss of blood through injury weakens not only the Liver and the blood but also the Kidneys and the essence. So weak people should not give blood. When we replenish the Liver we usually also replenish the kidney and vica versa because the essence (hormones) and the blood run together. Many tonic herbs nourish both Liver blood and kidney essence at the same time for example Nu Zhen Zi and He Shou Wu. We reinforce LIV.8 and KID.3 to tonify the Liver and Kidneys respectively.

Stomach and Liver

The Liver via the Gall Bladder is involved with digestion and the Liver and Stomach are both in the Middle Body Cavity (between the solar plexus and navel). So when a person has gastro-enteritis (inflammation of Stomach and intestines due to infection from bacteria/ virus/food poisoning) we not only clear and calm the Stomach and Large Intestine but also the Liver. The Liver would have become aggravated not only because of its close proximity to the Stomach but also because it will be trying to filter out the pathogens that have caused the gastro-enteritis. We reduce CV.12 to calm down excessive conditions of the MBC and apply moxa sticks to the handle of the needle.

Chapter 3

How to construct a Herbal Formula

3.1 Guidelines

After we have done a consultation and a TCM diagnosis we can create a herbal formula. With a herbal formula there are some addition general principles to consider.

- If the patient is too warm, we give them cool herbs
- If the patient is hot, we give them cold herbs
- If the patient is a bit cool / weak, we give them warm herbs
- If the patient is very cold / hypothermia, we give them hot herbs

- We **never** give a patient who is too hot herbs which are hot
- We **never** give a patient who is too cold herbs which are cold

- If the patient is sweating when they should not be, we give them astringent herbs
- If the patient is not sweating when they should be, we give them diaphoretic herbs

- We **never** give a patient who is sweating when they should not be a diaphoretic herb
- We **never** give a patient who is not sweating when they should be an astringent herb

- If a person has oedema we give them a diuretic herb to clear excess water
- If a person is dehydrated we give them moistening herbs

- We **never** give a patient who is dehydrated a diuretic herb
- We **never** give a patient who has oedema moistening herbs

- If a patient has low blood pressure we can prescribe lifting herbs
- If a patient has high blood pressure we can prescribe descending herbs

- We never give a patient who has high blood pressure lifting herbs
- We never give a patient who has low blood pressure descending herbs

There are many more examples that can be given but the principle remains the same, however when we are balancing our formula, it is neccecary to not be ruled one hundred percent by this principle.

3.2 Constructing and Balancing

First We explain how to construct a formula, then how to balance it.

3.2.1 Step 1 : Constructing

When we construct our herbal formula we:

1. Identify the main cause of the patients condition that needs attention and select herbs to treat it
2. Choose support herbs to further reinforce the action we are trying to achieve
3. Add herbs to treat the patients symptoms that are produced by the underlying cause of the problem

If appropriate, we add a guide herb to take the formula to the part of the body where it is needed most.

Next we balance our formula in relation to the patient and the formula itself.

3.2.2 Step 2 : Balancing

The balancing of a formula in relation to the patient means that if the patient had an acute excess condition like a high grade temperature, we would reduce it with a cold herb (Huang Qin), however to avoid the problem of the cold herb weakening the functional power of their bodies we would balance our formula in relation to the patient by adding a warming tonic herb (Bai Zhu) for the Spleen to keep their bodies immunity strong.

So in this case we are using a warm herb to balance a cold herb. The opposite is also true, if the patient had hypothermia we would give them a hot herb (Xian Mao) and to balance this we would give them a cool herb (Bai Shao Yao) to stop the hot herb cooking their blood and damaging it.

So we construct our herbal formula in a balanced way to bring health to the patient through returning their body back into a state of balance.

The construction of Acupuncture and Chinese Herbal Medicine formulas are both a science and an art, only through dedicated daily study and contemplation and through practice and experience over time can the correct process be understood.

As time passes our formulas become more comprehensive and effective and we are able to give a bigger, better benefit to the patient in a shorter time.

3.3 General Information about Treatments

When to take the Herbs

If the patient has a problem in the upper body, take herbs after having eaten. If the problem is in lower body, take herbs first and then eat. If the formula is mostly a tonic formula, it should be taken in the morning. If the formula is mostly sedating, it should be taken before going to sleep at night. If the condition is acute and an emergency, all herbs should be taken at once. If the condition is chronic, the herbs should be taken as small daily doses over the course of the week.

When not to treat a Patient

In certain circumstances we do not treat the patient:

- If they are are drunk we do not treat them
- Do not treat a pregnant woman unless you have had special training, the first three months of a pregnancy are very delicate, so even if you have had special training great care should be taken at this time
- If the patient has an acute life threatening condition that is best treated with Western medicine then refer them to their doctor or if it is more serious call an ambulance immediately

Chapter 4

Herbs used in Clinical Practice

There are approximately 500 herbs in TCM. At The College of Chinese Medicine we regularly use about 200 herbs in clinical practice with another 70 to 80 for rare conditions. Here is a list of the 272 herbs currently used at The College.

The page numbers are for the book:

Chinese Herbal Medicine
Materia Medica Revised Edition
by Dan Bensky and Andrew Gamble
ISBN: 0939616157

This book is an excellent reference work recommended to First Year students. When a student is made aware that a herb can be used for a particular condition they should look it up and read all about it at the soonest possible opportunity.

4.1 Herbs Reference List

4.1.1 Herbs that Release the Exterior

Table 4.1: Warm Acrid herbs

Properties	Name	Page
Warm acrid herbs	Ma Huang	p.28
	Gui Zhi	p.29
	Zi Su Ye	p.30
	Jing Jie	p.31
	Fang Feng	p.32
	Qiang Huo	p.33
	Gao Ben	p.34
	Bai Zhi	p.34
	Xi Xin	p.35
	Xin Yin Hua	p.39

Table 4.2: Cool Acrid herbs

Properties	Name	Page
Cool acrid herbs	Bo He	p.40
	Niu Bang Zi	p.41
	Chan Tui	p.42
	Sang Ye	p.43
	Ju Hua	p.44
	Fu Ping	p.46
	Chai Hu	p.49
	Sheng Ma	p.50

4.1.2 Herbs that clear Heat

Table 4.3: Herbs that drain fire

Properties	Name	Page
Herbs that drain fire	Shi Goa	p.55
	Zhi Zi	p.57
	Xia Ku Cao	p.59
	Lu Gen	p.61
	Mi Meng Hua	p.64

Table 4.4: Herbs that cool the blood

Properties	Name	Page
Herbs that cool the blood	Sheng Di Huang	p.68
	Xuan Shen	p.69
	Mu Dan Pi	p.70
	Zi Cao	p.71
	Di Gu Pi	p.72

Table 4.5: Herbs that clear heat and relieve toxicity

Properties	Name	Page
Herbs that clear heat and relieve toxicity	Jin Yin Hua	p.85
	Lian Qiao	p.86
	Qing Dai	p.88
	Ban Lan Gen	p.88
	Pu Gong Ying	p.89
	Zi Hua Di Ding	p.90
	Ye Ju Hua	p.91
	Bai Hua She She Cao	p.95
	Bai Tou Weng	p.97
	Bai Xian Pi	p.100
	Tu Fu Ling	p.101
	Ban Zhi Lian	p.102
	Shi Shang Bai	p.106

Table 4.6: Herbs that clear heat & dry dampness

Properties	Name	Page
Herbs that clear heat & dry dampness	Huang Qin	p.75
	Huang Lian	p.77

Table 4.7: Herbs that clear & relieve summerheat

Properties	Name	Page
Herbs that clear & relieve summerheat	Qing Hao	p.110

4.1.3 Downward Draining Herbs

Table 4.8: Purgative

Properties	Name	Page
Purgative	Da Huang	p.115
Herbs that drain dampness	Fu Ling	p.131
	Hua Shi	p.133
	Yi Yi Ren	p.134
	Dong Gua Ren	p.135
	Bei Xie	p.143
	Jin Qian Cao	p.144
	Yin Chen Hao	p.146

Table 4.9: Herbs that dispel wind-dampness

Properties	Name	Page
Herbs that dispel wind-dampness	Du Huo	p.155
	Qin Jiao	p.156
	Mu Gua	p.159
	Sang Zhi	p.161
	Wu Jia Pi	p.161
	Cang Er Zi	p.162
	Hai Feng Teng	p.165

4.1.4 Herbs that transform Phlegm & stop coughing

Table 4.10: Herbs that cool and transform phlegm-heat

Properties	Name	Page
Herbs that cool and transform phlegm-heat	Gua Lou (fruit)	p.178
	Gua Lou Ren (seed)	p.179
	Tian Hua Fen (root)	p.180
	Zhu Ru	p.183
	Kun Bu	p.185
	Hai Zao	p.185

Table 4.11: Herbs that relieve coughing and wheezing

Properties	Name	Page
Herbs that relieve coughing and wheezing	Bai Bu	p.202
	Sang Bai Pi	p.203

Table 4.12: Warm herbs that transform phlegm-cold

Properties	Name	Page
Warm herbs that transform phlegm-cold	Ban Xia	p.190
	Jie Geng	p.196
	Zao Jiao Ci	p.198

Table 4.13: Aromatic herbs that transform dampness

Properties	Name	Page
Aromatic herbs that transform dampness	Hou Po	p.215
	Cang Zhu	p.216
	Bai Dou Kou	p.217

4.1.5 Herbs that relieve food stagnation

Table 4.14: Herbs that relieve food stagnation

Properties	Name	Page
Herbs that relieve food stagnation	Shen Qu	p.226
	Ji Nei Jin	p.226

4.1.6 Herbs that regulate the *qi*

Table 4.15: Herbs that regulate the *qi*

Properties	Name	Page
Herbs that regulate the *qi*	Chen Pi	p.232
	Qing Pi	p.233
	Zhi Shi	p.234
	Xiang Fu	p.236
	Mu Xiang	p.237
	Xie Bai	p.240

4.1.7 Herbs that regulate the blood

Table 4.16: Herbs that stop bleeding

Properties	Name	Page
Herbs that stop bleeding	Pu Huang	p.249
	Xian He Cao	p.250
	San Qi	p.251
	Bai Ji	p.253
	Di Yu	p.255
	Huai Hua Mi	p.256

Table 4.17: Herbs that invigorate the blood

Properties	Name	Page
Herbs that invigorate the blood	Chuan Xiong	p.266
	Dan Shen (Miltiorrhizae)	p.267
	Ji Xue Teng	p.268
	Yan Hu Suo	p.269
	Yu Jin	p.271
	Yi Mu Cao	p.273
	Chi Shao	p.277
	Tao Ren	p.278
	Hong Hua	p.279
	E Zhu	p.280
	San Leng	p.281
	Ru Xiang	p.282
	Mo Yao	p.283
	Niu Yi	p.284
	Wang Bu Liu Zing	p.285
	Xue Jie	p.288
	Su Mu	p.288
	Shui Zhi	p.291

4.1.8 Herbs that warm the interior

Table 4.18: Herbs that warm the interior

Properties	Name	Page
Herbs that warm the interior & expel cold	Gan Jiang	p.300

4.1.9 Tonifying Herbs

Table 4.19: Herbs that tonify the *qi*

Properties	Name	Page
Herbs that tonify the *qi*	Dang Shen (Codonopsis)	p.317
	Huang Qi	p.318
	Shan Yao	p.320
	Bai Zhu	p.321
	Da Zao	p.323
	Gan Cao	p.323

Table 4.20: Herbs that tonify the *yang*

Properties	Name	Page
Herbs that tonify the *yang*	Yin Yang Huo	p.341
	Xian Mao	p.346
	Du Zhong	p.347[1]
	Gou Ji	p.348
	Xu Duan	p.349
	Gu Sui Bu	p.350
	Tu Si Zi	p.350

Table 4.21: Herbs that tonify the blood

Properties	Name	Page
Herbs that tonify the blood	Shu Di Huang	p.327
	He Shou Wu	p.328
	Dang Gui	p.329
	Bai Shao	p.331
	E Jiao	p.332
	Gou Qi Zi	p.333
	Sang Shen	p.334
	Long Yan Rou	p.335

Table 4.22: Herbs that tonify the *yin*

Properties	Name	Page
Herbs that tonify the *yin*	Tian Men Dong	p.359
	Mai Men Dong	p.360
	Shi Hu	p.361
	Bai He	p.363
	Sang Ji Sheng	p.364
	Han Lian Cao	p.365
	Nu Zhen Zi	p.365

Table 4.23: Herbs that stabilize and bind

Properties	Name	Page
Herbs that stabilize and bind	Shan Zhu Yu	p.375
	Wu Wei Zi	p.376
	Qian Shi	p.386
	Jin Ying Zi	p.387
	Fu Pen Zi	p.388
	Fu Xiao Mai	p.390
	Hai Piao Xiao	p.392

4.1.10 Substances that calm the spirit

Table 4.24: Substances that anchor, settle and calm the spirit

Properties	Name	Page
Substances that anchor, settle and calm the spirit	Long Gu	p.397
	Mu Li	p.398
	Zhen Zhu	p.402

Table 4.25: Herbs that nourish the Heart & calm the spirit

Properties	Name	Page
Herbs that nourish the Heart & calm the spirit	Suan Zao Ren	p.404
	Bai Zi Ren	p.405
	He Huan Pi	p.406
	Ye Jiao Teng	p.407

Table 4.26: Aromatic substances that open the orifices

Properties	Name	Page
Aromatic substances that open the orifices	Bing Pian	p.414
	Shi Chang Pu	p.415

Table 4.27: Substances that extinguish wind & stop tremors

Properties	Name	Page
Substances that extinguish wind & stop tremors	Gou Teng	p.423
	Tian Ma	p.424
	Di Long	p.426
	Quan Xie	p.427
	Wu Gong	p.428
	Jiang Can	p.429

Table 4.28: Substances for external application

Properties	Name	Page
Substances for external application	Ming Fan	p.447
	Xiong Huang	p.449
	Peng Sha	p.450
	Liu Huang	p.451
	Er Cha	p.455

4.2 Commonly used herbs in herbal formulas used in pairs to balance each other

When making formulas, certain herbs within the formula are used together to support each other or to balance side effects of other herbs. Here is a list of commonly used herbs and how to balance them in a formula. Also some basic formulas for common conditions are shown.

Bai Zhu and Fu Ling are often used together. Both strengthen the Spleen and digestive system and both dry up damp (Fu Ling is diuretic and Bai Zhu strengthens the metabolic process of the Spleen to dry damp in the body).

Bai Zhu lifts up and Fu Ling drains downwards. They can be used for oedema and bone swelling.

Fu Ling is an appetite suppressant (used in weight lose formula) and Bai Zhu is an appetite stimulant - given to post-op, or post illness patients in the recovery stages to boost their appetite. So, to strengthen the Spleen and digestive system without exciting or suppressing the appetite - use both together.

A formula to heal the Heart

- Xie Bai
- Gua Lou
- Dang Shen (Codonopsis)
- Ban Xia
- Zhu Ru
- Bai He

As Xie Bai weakens the *qi* we add the tonic Dang Shen (Codonopsis) to strengthen *qi*. Because Gua Lou and Bai He are moistening we add Ban Xia to evenly disperse the fluid. Zhu Ru will clear the heat and Dang Shen (Codonopsis) will keep the Heart strong whilst we do this.

Recovery formula, increase *qi*, boost/rebuild immune system and counter fatigue

- Bai Zhu
- Dang Gui
- Dang Shen (Codonopsis)
- Gan Cao
- Huang Qi.

To invigorate *qi* & blood circulation to counter blood stagnation causing pain

- Yu Jin (cools)
- Yan Hu Suo (warms, so they balance each other)

For even stronger effects of invigorating *qi* and blood circulation and thinning the blood to counter blood stagnation causing pain we use E Zhu and San Leng. For very, very serious blood stagnation, congestion, coagulation we can also add Shui Zhi (Leech). A mild blood moving herb is Tao Ren

Mu Dan Pi is used to clear heat from the blood. To keep the blood strong whilst we do this Shen Zhu Yu is used.

If using Du Zhong or Xian Mao which warm the kidney, then to counter this heat we can use Chi Shao Yao (cools blood in LBC) and Nu Zhen Zi which will nourish the essence so the heat does not burn it away. Also Gan Cao can be used to balance the slight toxic aspects of Xian Mao.

If there is a high grade temperature, Huang Qin is used. This herb has a very strong cold action which can weaken the Spleen and so it must be balanced with Bai Zhu or Dang Shen (Codonopsis) or Gan Cao. If there is an EPA with MBC congestion then we use Shen Qu.

Kun Bu (kelp) is often paired with Hai Zao (Sargassum Seaweed) for excess or deficiency of the thyroid gland because they both have a similar action.

For nasal congestion and inflamation, sinusitus and rhinoritis

- Xin Yi Hua
- Bai Zhi

For a drippy nose

- Bo He
- Bai Zhu
- Bai Zhi

To counter Spleen deficiency diarrhoea

- Shan Yao
- Qian Shi
- Bai Zhu
- Gan Cao
- Sheng Ma
- Chai Hu
- Huang Qi
- Dang Shen (Codonopsis)

To clear insect bite, nettle rash, summer heat rash – Use the *Fire Flower Formula*

- Ju Hua
- Pu Gong Ying
- Jin Yin Hua
- Lian Qiao
- Zi Hua Di Ding
- Di Ding
- We balance this formula with Gan Cao (tonic and detox)

Astringing Bladder to counter excess urination due to deficiency

- Fu Pen Zi
- Jin Ying Zi

To clear phlegm congestion of MBC and LBC

- Zhi Shi
- Zhi Ke
- Ban Xia
- Zhu Ru - if heat

For traumatic Injury

- Mo Yao
- Ru Xiang
- We balance with Bai Zhu and Gan Cao

Bai Zhu to help digest Mo Yao and Ru Xiang, and Gan Cao to detoxify

For anaemic blood

- Bai Zhu
- He Shou Wu
- Ji Xue Teng
- Sheng Di Huang
- Shu Di Huang
- Dang Gui
- Chuang Xiong
- Mu Dan Pi and Dan Shen (militiorrhizae) - if there is blood heat as well

For constipation caused by dryness of bowels

- Tao Ren
- Xing Ren
- Gua Lou Ren
- Huo Ma Ren
- Sang Shen
- Dang Gui
- Da Huang

To lift collapse/prolapse of Bladder, anus, GU system or Stomach

- Huang Qi
- Bai Zhu
- Gan Cao
- Dang Shen (Codonopsis)
- Chai Hu
- Sheng Ma
- Zhi Ke

For PMT or PMS

- Chai Hu
- Dang Gui
- Hong Hua
- He Huan Pi
- Pu Gong Ying - if swollen and tender breasts
- Yu Jin
- Yan Hu Suo
- Bai Shao Yao

Dang Gui and Hong Hua all move blood through the female GU System very, very strongly, regulating menstruation and stopping cramps. Prescribe in the week prior to menstruation.

For post menstrual blood loss, to increase blood volume

- Shu Di Huang
- Sheng Di Huang
- Bai Zhu - to metabolise these herbs. Also Bai Zhu helps the Spleen make blood.

For Wind Damp Attack to UBC

- Jing Jie
- Fang Feng (or Qiang Huo if its a strong attack)
- Du Huo (for the LBC).

These are all warm diaphoretics to clear wind, cold and damp.

For Heart Blood deficiency with panic

- Ye Jiao Teng
- He Huan Pi

Headache, dizziness & vertigo caused by wind attack to head with blood deficiency

- Jing Jie
- Chuan Xiong
- Bai Zhi

To detoxify the Liver and Blood

- Da Huang with any detox herb
- Ban Zhi Lian
- Zao Xiu
- Pu Gong Ying
- Lian Qiao
- Jin Yin Hua

To counter Anal Bleeding

- Di Yu Tan
- Zhi Zi Tan

E Jiao can be added but it will stop menstruation, so only prescribe to women three days after menstruation. Add Bai Zhu when using E Jiao to help metabolise it.

Stomach Ulcer

- Bai Ji
- Gan Cao
- Hai Piao Xiao

To stop cough

- Tao Ren
- Bai Shao Yao
- Gan Cao
- Bai He
- Bai Bu
- Ban Xia
- Da Huang

Essence tonification

- He Shou Wu
- Tian Men Dong
- Gou Qi Zi
- Zhi Mu
- Sang Shen Zi
- Shan Yao
- Nu Zhen Zi

Blood tonification

- Dang Gui
- He Shou Wu
- Shu Di Huang
- Bai Zhu
- Da Zao

***qi* tonification**

- Bai Zhu
- Da Zao
- Dang Shen (Codonopsis)
- Gan Cao
- Huang Qi
- Shan Yao
- Dang Gui

Kidney *yang* tonification

- Du Zhong
- Bu Gu Zhi
- Xian Mao
- Ba Ji Tian
- Yin Yang Huo

Blood stasis

- Chi Shao Yao
- Tao Ren
- Hong Hua
- San Leng
- E Zhu
- Shui Zhi
- Chuan Xiong
- Yan Hu Suo
- Yu Jin
- He Huan Pi

Stop bleeding

- Di Yu Tan*
- Zhi Zi Tan*
- Pu Huang Tan*
- Bai Mao Gen Tan*
- Bai Ji
- E Jiao

(* Tan means charcoaled)

Urinary heat & infection

- Tu Fu Ling
- Ze Xie
- Hua Shi
- Jin Yin Hua
- Lian Qiao
- Bai Mao Gen
- Mu Tong
- Zhi Mu

Indigestion with bloating & wind

- Hai Feng Teng
- Zhi Shi
- Shen Qu
- Ban Xia
- Yu Jin
- Yan Hu Suo
- Zhi Zi
- Hou Po
- Chen Pi

diarrhoea & dysentry

- Gan Cao
- Bai Zhu
- Da Huang
- Bai Tou Weng
- Bai Hua She She Cao
- Shan Yao

Intercostal pain & Liver *qi* stagnation

- Xiang Fu
- Chuan Lian Zi
- Yu Jin
- Chai Hu
- Zhi Zi
- Ju Hua
- Qing Pi

Heat & fever

- Qing Dai
- Da Qing Ye
- Ban Lan Gen
- Gou Teng
- Nu Zhen Zi
- Zhi Mu

Astringent

- Wu Wei Zi
- Shan Yao
- Hai Piao Xiao
- Jin Ying Zi
- Fu Pen Zi
- Fu Xiao Mai
- Huang Qi

Oedema of the legs

- Sang Bai Pi
- Kun Bu
- Yi Yi Ren
- Wu Jia Pi

- Fu Ling Pi
- Bai Zhu
- Ze Xie

Detoxifying seafood poisoning

- Jing Jie
- Zi Su Ye
- Sheng Jiang
- Gan Cao
- Da Huang
- Pu Gong Ying

To Strengthen the defensive *qi* & improve the bodies resistance to Wind Attack

- Bai Zhu
- Huang Qi
- Fang Feng

To clear headaches

- Jing Jie - back of head
- Gao Ben - top of head
- Bai Zhi - front of head
- Chuan Xiong - headache from blood deficiency

Chapter 5

How to construct an acupuncture formula

After we have done a consultation and a TCM diagnosis we can create an acupuncture formula. We first have to decide if we are treating an internal or external medicine condition.

Internal/External

External medicine is also called musculoskeletal medicine, this is sports injuries/traumatic injuries. Internal medicine is general health problems caused by imbalances in the processes controlled by the internal organs.

Often a patient will have a combination of both internal and external medicine mixed together.

Excess/deficiency

Next we have to decide if the disease they are suffering from is excessive or deficient.

Often a patient will have a combination of both excessive and deficient conditions mixed together.

We then construct an acupuncture formula, if it is an excessive condition we reduce the acupuncture points on the extremities first, if it is a deficiency condition we reinforce the acupuncture points on the torso first.

Combining Local and Distal Points

If the patient has a problem on a certain location we acupuncture this location, whether it is an internal or external medicine problem.

For example, if they have an inflamed knee due to a fall we stick needles in the knee, if they have an inflamed abdomen due to food poisoning, we stick needles in the abdomen.

We then support our work with what are called adjacent points and then we further expand our formula with the distant points, these are acupuncture points on the extremities, the wrists/hands and ankles/feet.

So for example if a patient had bloating, pain, spasms, swelling and inflammation of the MBC due to food poisoning we would choose the point CV.12 reducing, this point can be supported with adjacent points such as CV.13, KID.19, and ST.21. We would then add the distal points such as LI.4, ST.36, ST.37, ST.44, GB.34, LIV.3 and SP.6 all reducing. So we would be reducing the excessive condition locally and also draining off the excess from the ends of the meridians on the extremities.

We also often begin constructing our formula by starting with small formulas such as the eight trigrams formulas and the additional paired points formulas and then adding to them or combining them. Our formula then expands until we have covered all the areas that need attention.

Also at CCM we refer to certain points that are grouped together as the *yin System* and another group of points as the *yang System*. Using this concept we have another way of constructing our acupuncture formulas.

The *yin* System is KID.3 and SP.6 (we sometimes add KID.6, SP.9 and SP.10). The *yang* System is LIV.3, BL.60 and GB.39 and GB.40 (we sometimes add ST.36 and GB.34). So for example if the patient had heat and was deficient we would reduce the *yang* System to clear the heat and reinforce the *yin* System to replenish their deficiency.

We also have to keep in mind that each point we are using when constructing a formula has to be balanced.

For example we balance KID.3 reinforcing with BL.60 reducing.

We balance SP.6 reducing with BL.20 and BL.23 reinforcing.

Also when constructing an acupuncture formula we will be using points that will be treating the cause of the patients condition and we will also be using points that treat the symptoms produced by that condition. We will also be treating the patients underlying general health condition.

How to avoid pneumothorax

There is a medical condition called pneumothorax which means a puncture of the thoracic cavity, in serious cases this results in air entering into the thoracic cavity and the collapse of the lung. If pneumothorax was caused by an acupuncture needle and the patient went to hospital they would be put under observation and no action would be taken because the hole was so small and the body would heal itself.

It is only when there is a major traumatic injury that a procedure is used. Nevertheless to avoid the occurance of pneumothorax we always insert the needles at an angle of about 25 degrees when acupuncturing on the back above the level of thoracic vertebra 12 (T 12).

When acupuncturing on the front of the torso it is the same, so when the most used points, LU.1 and KID.27 are acupunctured the needles are inserted at an angle of about 25 degrees. However for CV 17 the needle is almost flat against the skin, this is for two reasons, first the sternum (breastbone) is just under this point and also because there is a rare condition called sternal foramen (a hole in the sternum) and to avoid going through this in to the thoracic cavity and causing pneumothorax we have a very flat angle of insertion for the acupuncture needle.

5.0.1 How to avoid the patient fainting

Never acupuncture a patient who is sitting up, only acupuncture a patient when they are lying down. This will significantly reduce the possibility of them fainting because gravity will keep blood going to the head. If the patient was sitting up, the blood could easily drain from the head.

5.0.2 How to revive consciousness

There are two types of collapse where a person loses consciousness: prostrational collapse and occlusional collapse.

1. Prostrational collapse: the patients eyes, mouth and hands will be open, they have a deficiency condition
2. Occlusional collapse: is very rare, the patients eyes will be shut and their hands and teeth clenched, they have an excessive condition

Prostrational Collapse - what to do

In this situation to revive/restore consciousness we put them on their back and elevate the legs and acupuncture GV.26

Then massage up each side of the back of the neck from SI.15 towards BL.10 & GB.20 to maintain consciousness, and we acupuncture in the reinforcing direction, KID.3, SP.6, ST.36, LU.1 and KID.27 to strengthen the whole system.

When they are conscious and sitting up again we give them a warm, sweet drink.

Occlusional collapse - what to do

To revive/restore consciousness we punch and squeeze a drop of blood (SB) from the ends of all fingers and thumbs. This has a releasing and relaxing effect. If you cannot get to the finger tips (because the fists are clenched too tightly), then do the toes.

Also we punch and squeeze blood from the apex of the ear.

5.0.3 How to avoid causing the patient discomfort when inserting acupuncture needles

There are two movements involved when inserting the needle, the first is the initial punch which enables the point of the needle to quickly go through the skin surface, the second is when the needle is threaded to the correct depth.

When we punch it is a very focused and quick technique so the patient does not have the nerve ends on the skin surface activated, so there is not really any discomfort.

Threading is a slower technique, to avoid having the patient experience unnecessary discomfort we do not insert the needle with the slow threading technique. This would just activate all the nerve endings on the surface and cause the patient discomfort and pain.

So when inserting the needle, **PUNCH. Do not thread!**

5.0.4 Needle direction on meridians

Each meridian flows in one direction, if we insert the needle in the direction that the *qi* energy flows we will have an increasing effect, we use this to strengthen deficiency conditions (weakness, poor functional power, immune system collapse). If we insert the needle against the direction of the *qi* flow in a meridian we are reducing, this is to counter excessive conditions (pain, swelling, inflammation, stagnation/congestion).

We would not put the needles into a meridian very close to each other in opposite directions, this would confuse the body. However there are several situations where we do put the needles in opposite directions in the same meridian but they are at different ends of the meridian and so because they are far enough away from each other there is no problem.

Here are the main combinations:

We can do points on the back of the body on the Shu points of the Bladder meridian reinforcing and at the same time do points on the Bladder meridian on the legs reducing.

We can do Lung 1 reinforcing on the chest and lung points on the arm reducing.

We can reinforce Spleen points on the legs and reduce Spleen points on the abdomen.

5.1 Proportional Body Measurements

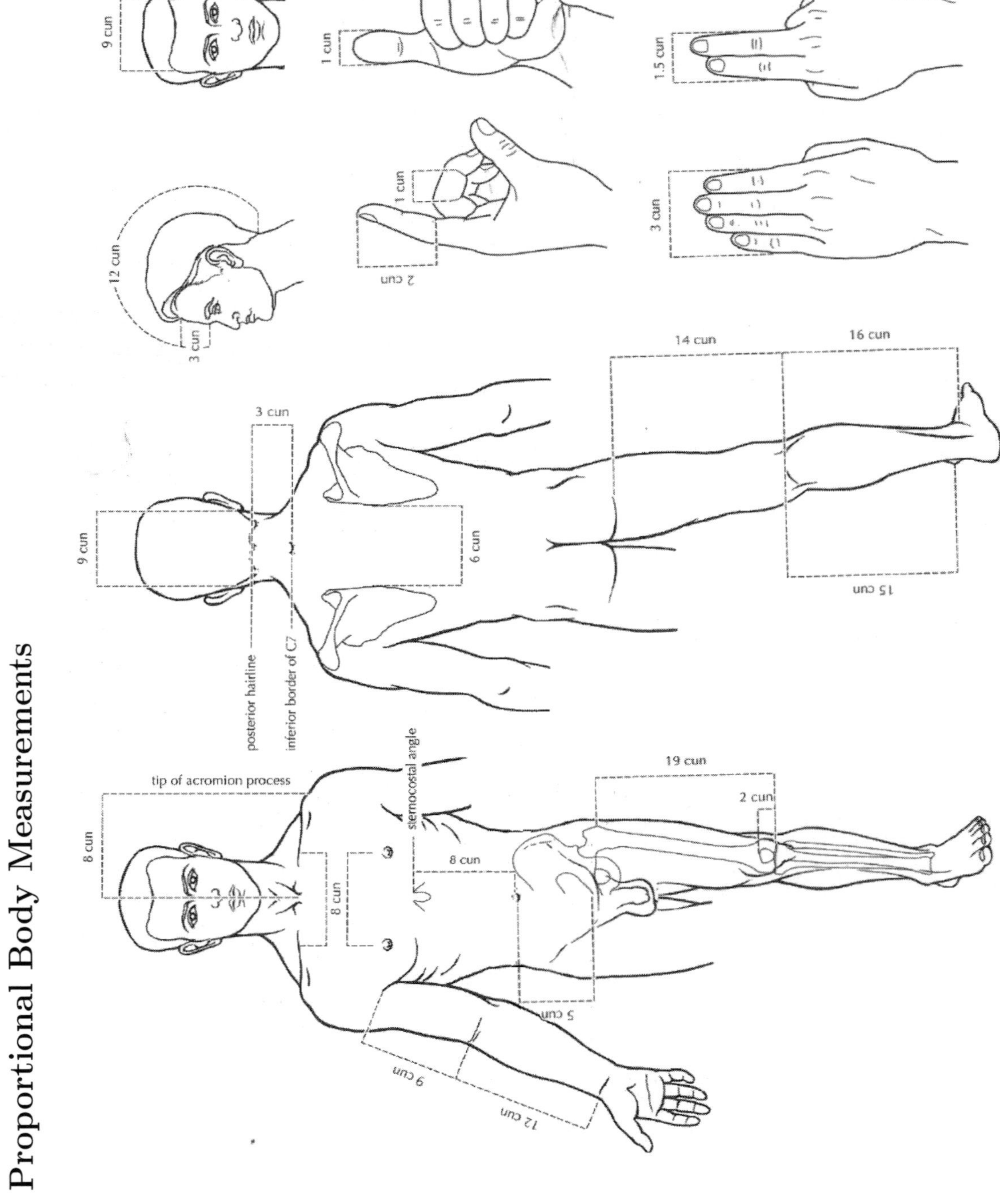

Chapter 6

Acupuncture points used in clinical practice

The arrow shows the direction of the *qi* energy flow in that meridian, to reinforce insert the needle in the direction of the *qi* flow and to reduce we go against the direction of the *qi* flow. So the needle will always be at an angle, however there are one or two points that have to be acupunctured vertically for anatomical reasons.

6.1 The Lung Meridian

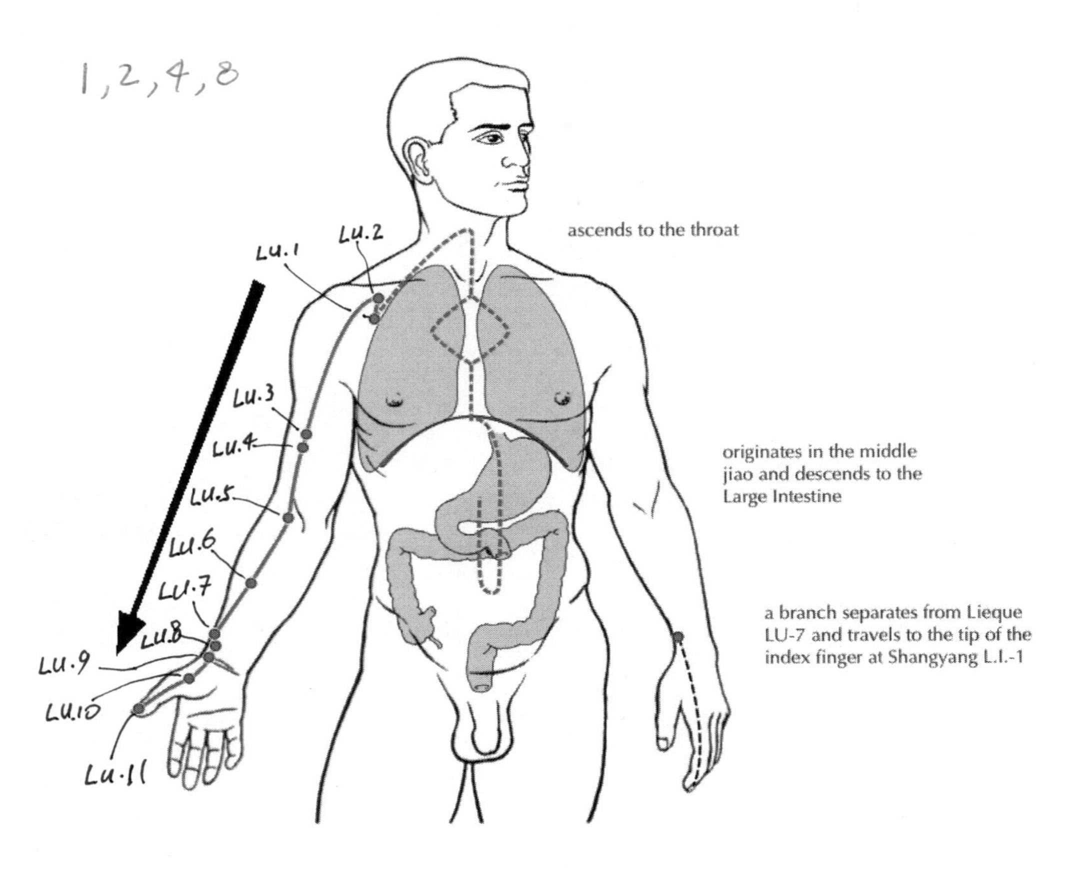

The Lung primary channel

Discussion

The Lungs are situated in the the thoracic cavity which is part of the Upper Body Cavity (UBC), they are the first major internal organs to come into contact with the environment and so are vulnerable to Exogenous Pathogenic Attack (EPA). They must be kept strong and be able to adapt to the changing weather of the seasons. People often succumb to an Exogenous Pathogenic Attack when the weather changes suddenly or during the change of the seasons. The first sign of this is often discomfort in the nose and throat, these are connected to the Lungs.

The Lungs extend to become the throat and nose and the skin is known as the third lung. The skin, nose and throat are the bodies first line of defence when Exogenous Pathogenic Factors attack the body. The pores of the skin open when there is hot weather to release the heat and they close in cold weather to keep the heat in. The Lungs control respiration and the skin also breathes.

If a patient has excess heat/temprature from an EPA and they are not sweating we can make them sweat with a diaphoretic herb such as Jing Jie.

If they are sweating due to deficiency we can astringe them with Fu Xiao Mai. Also the acupuncture points LI.4 reducing combined with KID.7 reinforcing will stop sweating.

The Lungs through the process of sweating help to regulate the water metabolism in the body. So if a patient had eodema in the upper body one of the herbs in the formula would be a diaphoretic (e.g. Jing Jie).

Through the Lungs clean external air and energy are inhaled and waste energy and air from within are exhaled. If the Lungs ability to gather and mobilise *qi* is strong the *qi* of the whole body will be strong, if the Lungs become weak the *qi* of the whole body is affected. *Qi* is formed in the Lungs and goes from there to defend the bodies surface and is then called wei *qi* - defensive *qi*. The herb Huang Qi is used to strengthen the Lungs and the defensive energy.

If the Lungs are weak there may be general lassitude, feeble speech, weak respiration and shortness of breath. The Exogenous Pathogenic Factors that can affect the Lungs are viruses, bacteria, wind, cold, heat and dampness. The Lungs particularly dislike dryness which tends to occur more in the dry autumn air. If the Lungs are dry the skin goes dry, to moisten the skin, clear lung heat and moisten the Lungs we use Bai He and Lu Gen.

If the nose is attacked by Exogenous Pathogenic Factors there may be a runny nose, for this we use Bo He, if there is sinus/nasal congestion we use Xin Yi. If the throat is attacked there may be sore throat, dry throat or hoarse voice, for this we use Jin Yin Hua and Gan Cao. Whenever there is a problem with the skin, nose or throat, treat the skin, nose and throat but also treat the Lungs whose *qi* affects all these systems.

The taste which most affects the Lungs is pungent, this means that herbs with a pungent aroma will cause diaphoresis (sweating). This opening of the pores of the skin, the third lung, can help to clear pathogenic factors like wind, heat and damp from the system. To clear wind cold, a warm pungent herb would be used such as Jing Jie and to clear wind heat a cool pungent herb would be used such as Ju Hua.

6.1.1 The Lung Meridian points used in clinical practice

LU.1

Length of needle used : 1"
Depth of insertion : 0.5"
Location : On the lateral chest, inferior to the acromial end of the clavicle, 6" lateral to the Conception Vessel.
Action : Only used reinforcing to strengthens the Lungs and can also strengthen the Heart, this point has a shallow angle of insertion, about 30 degrees to avoid pneumothorax (puncture of the thoracic cavity). Moxa sticks can be applied to the handle of the needle and birds pecking technique can be used to strengthen weak Lungs and for more severe deficiency cases three moxa cones (3 Δ) can be used directly on this acupuncture point after the needle has been removed. This point is often combined with KID.27 reinforcing.

LU.5

Length of needle used : 1.5"
Depth of insertion : 1"
Location : At the elbow, in the cubital crease, in the depression lateral to biceps brachii tendon.
Action : Only used reducing for lung/chest pain and lung spasms and cough. No moxa used on this point. This point is often combined with LI.11 reducing.

LU.7

Length of needle used : 1.5"
Depth of insertion : 1"
Location : On the radial side of the forearm, 1.5" superior to the tip of the radial styloid process.
Action : Only used reducing, for throat and nose problems, EPA and headache. We often use KID.6 reinforcing with LU.7 reducing, they are paired trigram points. So if we where using LU.7 reducing to clear a sore throat, headache, EPA we would support this with KID.6 reinforcing. No moxa used on this point.

LU.10

Length of needle used : 0.5"
Depth of insertion : 0.3"
Location : Midpoint on the thenar eminence, on the dorsal-palmar surface.
Action : Only used reducing to clear heat from the Lungs and skin. No moxa used on this point. This point is often combined with LI.11.

LU.11

Length of needle used : 0.5"
Depth of insertion : 0.2"

Location : On the radial side of the thenar eminence, 0.1" posterior to the nailbed.
Action : Used reducing for arthritis or pain or inflammation or swelling along the thumb. For this moxa can be applied to the handle of the needle. We SB to clear UBC heat / temperature from virus / EPA, sore / swollen throat, respiratory failure, epilepsy and UBC heat or excess *yang* or HBP, when used in this way this point is often combined with SB LI.1 and SB on *Er Jian* (extra point on the apex of the ear).

6.1.2 Lungs herbs

Herb	**Actions**
Zhi Mu, Bai He, Shi Hu, Tian Men Dong, Lu Gen	To moisten Lungs to clear Lung dryness. HEAT CAUSES DRYNESS
Bai Bu	To stop coughing.
Jie Geng	To clear heat and phlegm or cold and phlegm and to conduct to the Lungs.
Hou Po	To clear cold and phlegm, to open the chest and to send Lungs *qi* downwards.
Zhu Ru, Lu Gen, Ban Xia	To clear thick congealed phlegm on the Lungs with heat.
Xi Xin, Ma Huang	To clear thin watery phlegm on the Lungs.
Xin Yi Hua, Bai Zhi	To clear thick congealed and congested nose.
Bo He, Bai Zhi	To clear thin watery nasal discharge.
Dang Shen (Codonopsis), Huang Qi 黄芪	To strengthen weak Lungs.
Jing Jie	To ventilate Lungs and release EPA from UBC by diaphoresis. A warm pungent herb if there is wind-cold.
Ju Hua	A cool pungent herb if there is wind-heat.
Sang Ye	A cold sweet herb if there is wind-heat.
Ju Hua, Sang Ye	Good for hayfever.

6.2 The Large Intestine Meridian

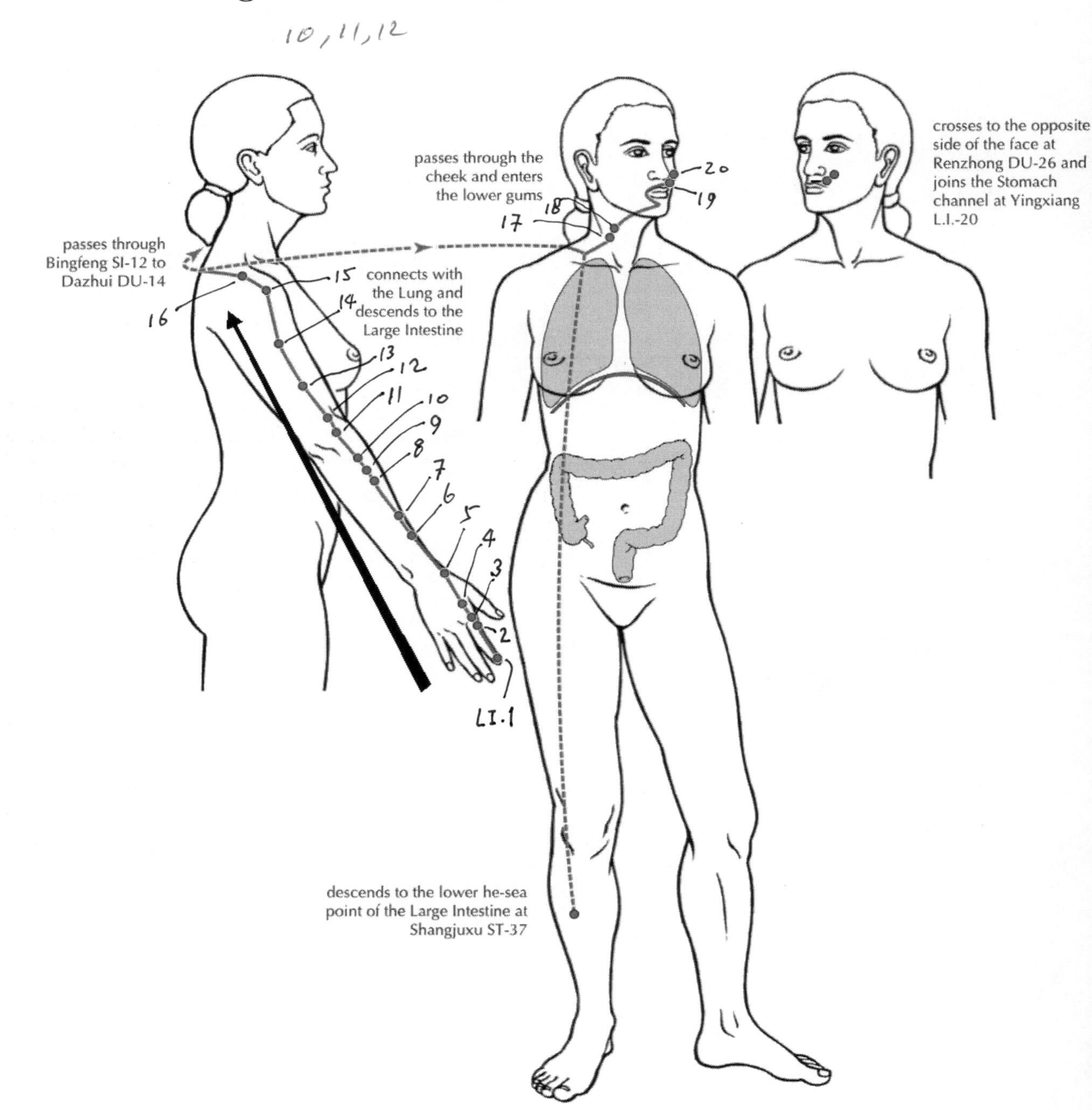

The Large Intestine primary channel

Illustrations reprinted from A Manual of Acupuncture by Peter Deadman & Mazin Al-Khafaji with Kevin Baker by kind permission of the publishers, www.jcm.co.uk.

Discussion

The Large Intestine is in the Lower Body Cavity (LBC), its upper end connects with the Small Intestine via the ileocecum and its lower end is the anus. The Large Intestine receives waste sent down from the Small Intestine. It passes the water to the Bladder to be cleared through urination and the Large Intestine then excretes solid waste as faeces. If there is heat in the Large Intestine it could cause constipation. If a person has a very weak health constitution there could be diarrhoea.

6.2.1 The Large Intestine Meridian points used in clinical practice

LI.1

Length of needle used : 0.5”
Depth of insertion : 0.2”
Location : On the radial side of the second digit, 0.1” from the corner of the nail bed.
Action : Used reducing for arthritis or pain, inflammation or swelling along index finger. For this moxa can be applied to the handle of the needle. We SB to clear UBC heat / temprature from virus / EPA, headache, inflammation in the face and neck and UBC heat or excess *yang* or HBP, when used in this way this point is often combined with SB LU.11 and SB on *Er Jian*, the Extra Point on the apex of the ear.

LI.4

Length of needle used : 1”
Depth of insertion : 0.6”
Location : On the dorsum of the hand, approximately at the midpoint of the second metacarpal bone, in the belly of the first interosseus dorsalis muscle.
Action : This is the feature point for headaches. Used reducing for pain, swelling and inflammation of the head, nose, throat and face, for these problems it is often combined with LU.7 reducing. When used for reducing pain, swelling and inflammation of the intestines LI.4 is often combined with ST.36 reducing (LI.4 is combined with many other points for a wide variety of conditions, these combinations are known as The Additional Paired Points).

LI.5

Length of needle used : 1”
Depth of insertion : 0.5”
Location : On the radial side of the wrist, distal to the tip of the radial styloid process, in the depression between the tendons of extensor pollicis longus and brevis, in the “anatomical snuff box”.
Action : Used reducing for arthritis, pain, inflammation, swelling of wrist. Moxa can be applied to the handle of the needle.

LI.10

Length of needle used : 1.5"
Depth of insertion : 1"
Location : On the radial side of the posterior antebrachial region, 2" distal to the cubital crease.
Action : Used reducing for arthritis, pain and swelling of the elbow. Moxa can be applied to the handle of the needle.

LI.11

Length of needle used : 1.5"
Depth of insertion : 1"
Location : On the lateral side of the cubital crease when the elbow is close to full flexion.
Action : Used reducing for lung and UBC heat and skin rash / itching and arthritis, pain and swelling of the elbow. It is often combined with LU.5. Moxa can be applied to the handle of the needle.

LI.12

Length of needle used : 1.5"
Depth of insertion : 1"
Location : On the lateral side of the cubital crease, 1 superior to LI.11, at the junction of the lateral supracondylar ridge of the humerus with the epicondyle. Locate LI.12 with the elbow flexed.
Action : Used reducing for arthritis, pain and swelling of the elbow. Moxa can be applied to the handle of the needle.

LI.14

Length of needle used : 1.5"
Depth of insertion : 1"
Location : On the lateral brachial region, on the anterior margin of the insertion of the deltoid muscle, on the line connecting LI.11 at the cubital crease and LI.15 inferior to the acromion.
Action : Used reducing for arthritis, pain, inflammation, swelling of shoulder. Used reinforcing for weakness of the shoulder joint. Moxa can be applied to the handle of the needle.

LI.15

Length of needle used : 2.5"
Depth of insertion : 2"
Location : On the shoulder in the depression on the anterior border of the acromial part of the deltoid muscle when the arm is abducted.
Action : Used reducing for arthritis, pain, inflammation, swelling of shoulder and goitre

(swelling in front of neck due to enlarged thyroid) and lymphodenitis (swelling of lymph nodes). Used reinforcing for weakness of the shoulder joint, however the direction of insertion remains the same (pointing down towards the elbow) for anatomical reasons, the reinforcing is achieved through the setting on the acupuncture autoscope machine and our needle manipulation technique. Moxa can be applied to the handle of the needle.

LI.16

Length of needle used : 1"
Depth of insertion : 0.5"
Location : On the superior aspect of the scapular region, in the depression posterior to the acromial extremity of the clavicle and anterior to the scapular spine.
Action : Used reducing for arthritis, pain, inflammation, swelling of shoulder. Used reinforcing for weakness of the shoulder joint. Moxa can be applied to the handle of the needle.

LI.18

Length of needle used : 1"
Depth of insertion : 0.5"
Location : On the neck, on the sternocleidomastoid muscle, 3" lateral to the laryngeal prominence.
Action : Used reducing for pain, inflammation, swelling of the neck and throat, goitre, laryngitis and pharyngitis, lymphodenitis. This point is often combined with ST.9. Moxa can be applied to the handle of the needle.

LI.20

Length of needle used : 1.5"
Depth of insertion : 0.5"
Location : In the nasolabial groove, 0.5" lateral to the nostril. For anatomical reasons this point is always acupunctured in the same direction, towards the bridge of the nose.
Action : Used to reduce pain, inflammation and swelling of the nose, it opens the nasal passages to clear congestion, to treat anosmia (loss of sense of smell) and rhinitis (inflammation of the mucus membrane of the nose), sinusitis and hay fever. We have a Length of needle used 1.5" and a Depth needle inserted to 0.5" because we need to have the needle away from the face when we apply moxa sticks to the end of the needle. We often support our use of LI.20 with the three extra points, Bi Tong (above LI.20), Yin Tang (3rd eye point between eyebrows) and the high point on the bridge of the nose, all with Length of needle used = 0.5 and Depth needle inserted to = 0.3"

6.2.2 Local Extra Points

Bi Tong

Bi Tong is punctured like LI.20 and so goes in one direction only. These points are all local, the distal points to support this work would be LU.7 reducing, this is because the nose is the operning of the lung. Also LI.4 reducing because it controls the head and face.

Yin Tang & Bridge of the nose

Yin Tang and the high point of the bridge of the nose can be acupunctured reinforcing to build up the resistance of the nose to EPA prior to the hay fever season and for anosmia, or they can be inserted in the reducing direction to reduce excessive conditions like pain, inflammation and swelling of the nose and nasal congestion.

6.2.3 The additional paired points - Large Intestine

These are small acupuncture formulas using LI.4 that are made extensive use of in clinical practice.

- **LI.4 Reinforced, SP.6 Reduced :** This combination will create a downwards movement of *qi* in the body. Used for amenorrhoea (no period), to induce overdue birth, and for endometriosis.
- **LI.4 Reduced, SP.6 Reinforced :** This combination will create a upwards movement of *qi* in the body. Used for deficiency patients who are weak with low immunity, light headed, fainting, dizziness and vertigo due to blood deficiency in the head. Also for people with organ collapse (visceral prolapse), it could be the Stomach, intestines or anus or sexual organs. These combination of points can also be used for hypermenorreah (excessive menstrual bleeding) due to deficiency we also use a prismatic needle punch and squeeze blood on SP.1 and LIV.1 on both big toes and burn three moxa jong on these points.
- **LI.4 Reduced, KID.7 Reinforced :** Used to stop excessive sweating due to deficiency.
- **LI.4 Reduced, ST.36 Reduced :** Used for indigestion, abdominal distention / bloating / swelling or food stagnation or food poisoning. If severe diarrohea, add ST.37. If very severe abdominal swelling and pain, add SP.6 reducing.

6.2.4 Large Intestine herbs

Herb	**Actions**
Tao Ren, Sang Shen, Huo Ma Ren, Gua Lou Ren, Dang Gui, Mai Men Dong, Tian Men dong, Da Huang (small dose)	Dryness of the Large Intestine causing constipation
Bai Hua She She Cao, Pu Gong Ying, Jin Yin Hua, Da Huang, Gan Cao	Purge Large Intestine due to viral / bacterial EPA, skin condition toxin, food poisoning, heat from Large Intestine
Di Yu (charcoaled), Zhi Zi, (charcoaled), Huai Hua Mi, Pu Huang, Da Huang (small dose) and Bai Mao Gen	To stop bleeding from the anus (the end of the Large Intestine).
Fang Feng, Hai Feng Tang	To clear Wind from the Large Intestine.

6.3 The Stomach Meridian

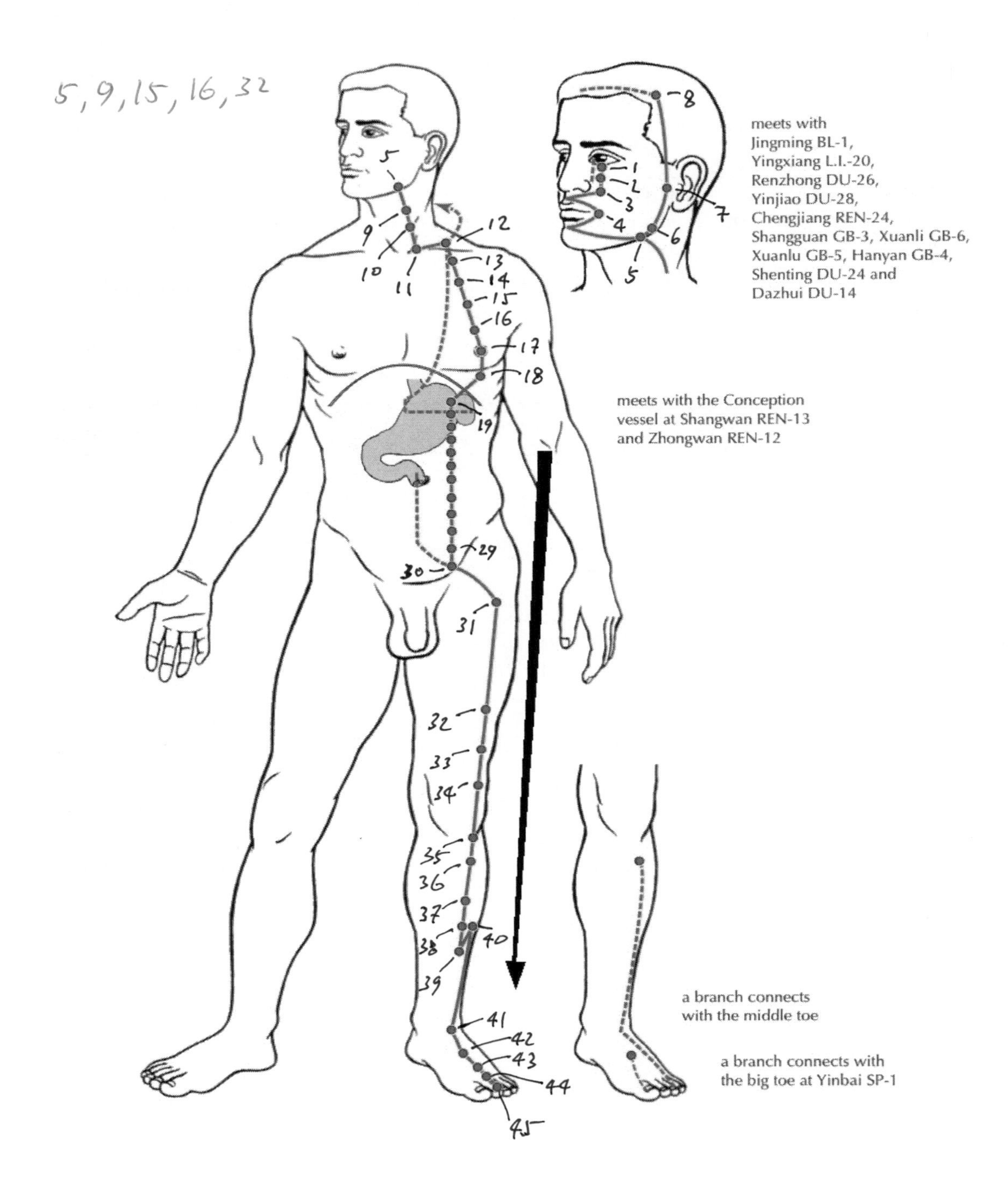

Illustrations reprinted from A Manual of Acupuncture by Peter Deadman & Mazin Al-Khafaji with Kevin Baker by kind permission of the publishers, www.jcm.co.uk.

Discussion

The Stomach is located in the Middle Body Cavity. It receives the food via the oesophagus and digests it and then passes on the remainder to the Small Intestine. The good nutrients from the food are extracted in the Stomach and given to the Spleen which absorbs them and uses them to make the blood and *qi* and to nourish the whole body.

The direction of the flow of the Stomach *qi* is down, if there is a reversal caused by a wind attack, viral or bacterial infection or drug intoxication it could result in nausea or vomiting. This is due to the body forming phlegm in an attempt to expel the pathogens from the system.

The mouth is the top end of the Stomach, so toxic food irritating the Stomach will cause mouth ulcers. If there is heat in the Stomach there could be pain, a burning sensation, unusual hunger, bleeding of the gums, constipation, and halitosis (bad breath).

When people have anxiety (butterflies in the Stomach) they want to eat sweet things to alleviate their anxiety. Extreme anxiety can result in Stomach ulcers. Sweet is the taste of the Spleen and Stomach and sweet tasting herbs and foods are used to strengthen the Spleen and Stomach but too much may harm them.

6.3.1 The Stomach points used in clinical practice

ST.3

Length of needle used : 1.5"
Depth of insertion : 0.5"
Location : On the face, level with the border of the ala nasi, in line with the pupil when the eyes are focused forward. For anatomical reasons this point is always acupunctured in the same direction, vertically.
Action : Used to reduce pain, inflammation and swelling of the face, upper jaw pain, facial and trigeminal neuralgia and numbness. We use Length of needle used = 1.5" and Depth needle inserted to = 0.5" because we need to have the needle away from the face when we apply moxa sticks to the end of the needle. We ask the patient to turn their head sideways before we apply moxa sticks to the handle of the needle.

ST.4

Length of needle used : 3"
Depth of insertion : 2.0"
Location : On the face, in the oral region, 0.5" lateral to the corner of the mouth in the first smile crease, in line with the pupil when the eyes are focused forward. We thread the needle through the cheek from ST.4 to ST.6, which is 1" from the corner of the jaw.
Action : Used for lower tooth and jaw ache and pain, grinding of teeth, facial paralysis, facial stroke. Moxa can be applied to the handle of the needle. We ask the patient to turn their head sideways before we apply moxa sticks to the handle of the needle.

ST.9

Length of needle used : 1"
Depth of insertion : 0.5"
Location : On the neck, on the anterior border of sternocleidomastoid muscle, at the level of the laryngeal prominence, 1.5" beside the Conception Vessel.
Action : Used reducing for pain, inflammation, swelling of the neck and throat, goitre, laryngitis, pharyngitis, tonsilitis, thyroiditis and HBP (high blood pressure) This point is often combined with LI.18. Moxa can be applied to the handle of the needle.

ST.21

Length of needle used : 1.5"
Depth of insertion : 1"
Location : On the abdomen, 2" lateral to the anterior midline, 4" superior to the umbilicus, at the level of CV.12.
Action : Used reducing for pain, inflammation, swelling of the MBC relating to the digestive system. This point is often combined with CV.12 and KID.19 reducing. Moxa can be applied to the handle of the needle.

ST.25

Length of needle used : 1.5"
Depth of insertion : 1"
Location : On the abdomen, 2" lateral to the umbilicus.
Action : Used reducing for pain, inflammation, swelling of the MBC and LBC relating to the digestive system. This point is often combined with ST.26 and SP.15. Moxa can be applied to the handle of the needle.

ST.26

Length of needle used : 1.5"
Depth of insertion : 1"
Location : On the lower abdomen, 1" inferior to the umbilicus and 2" lateral to the anterior midline, at the level of CV.7. 1" below ST.25.
Action : Used reducing for pain, inflammation, swelling of the MBC and LBC relating to the digestive system. This point is often combined with ST.25 and SP.15 reducing. Moxa can be applied to the handle of the needle.

ST.28

Length of needle used : 1.0"–1.5"
Depth of insertion : 0.5"–1.0"
Location : On the lower abdomen, 3" inferior to the umbilicus and 2" lateral to the anterior midline, at the level of CV.4.
Action : Used reducing for pain, inflammation, swelling of the LBC relating to the digestive system, the reproductive system or the genital urinary system. Can also

be used reinforcing to strengthen the reproductive system to counter impotence and infertility. This point is often combined with ST.29 reinforcing. Moxa can be applied to the handle of the needle

ST.29

Length of needle used : 1.0"–1.5"
Depth of insertion : 0.5"–1.0"
Location : On the lower abdomen, 1" above the pubic symphysis and 2" lateral to the anterior midline, at the level of CV 3 and 1" below ST.28.
Action : Used reducing for pain, inflammation, swelling of the LBC relating to the digestive system, the reproductive system or the genital urinary system. Can also be used reinforcing to strengthen the reproductive system to counter impotence and infertility. This point is often combined with ST.28 and the extra point, Zi Gong Xue.

ST.35

Length of needle used : 1.5"
Depth of insertion : 1.2"
Location : At the knee, in the depression known as 'the eyes of the knee', below the lateral side of the patella when the knee is flexed slightly.
Action : Used reducing for pain, inflammation, swelling of the knee joint. Can be used reinforcing to strengthen the knee joint. This point is often combined with the extra points He Ding and Xi Yan. Moxa can be applied to the handle of the needles.

ST.36

Length of needle used : 1.5"
Depth of insertion : 1"
Location : On the leg, one muscle strand lateral to the tibia's anterior crest, 3.5" inferior to ST.35.
Action : Used reducing for pain, inflammation, swelling of the MBC relating to the Stomach, Spleen and digestive system. The abdominal swelling could be caused by wind causing abdominal bloating or food stagnation or food poisoning causing inflammation and EPA attacking the MBC, gastritis, gastroenteritis, Stomach flu. For these excessive conditions this point is often combined with CV.12, CV.13, KID.19, ST.21, ST.25, ST.26, ST.28, LI.4, ST.37, ST.44, GB.34 and LIV.3. This point can be used reinforcing to strengthen the acquired *qi*, for this we would also reinforce SP.6 and KID.3, LU.1 and KID.27. Moxa can be applied to the handle of the needle for both excessive and deficient conditions and for more severe deficiency cases three moxa cones can be used directly on this acupuncture point after the needle has been removed. Traditionally people who are well can apply 3 Δ to this point every other day as a preventative measure to keep the body strong. There is also an additional obscure rare use for this point, if the patient has a shoulder problem around the area of LI.16 then ST.36 reducing can be a distal support point.

ST.37

Length of needle used : 1.5"
Depth of insertion : 1"
Location : On the leg, one muscle strand lateral to the tibia's anterior crest, 6.5" inferior to ST.35.
Action : This point is used to support ST.36 when it is done reducing for pain, inflammation, swelling of the MBC relating to the Stomach, Spleen and digestive system. It is most often used to reduce diarrhoea in cases of food poisoning and EPA attacking the MBC, gastritis, gastroenteritis, Stomach flu. Moxa can be applied to the handle of the needle.

ST.40

Length of needle used : 1"
Depth of insertion : 0.5"
Location : On the leg, 1" lateral to ST.38 at the midpoint of a line between ST.35 at the lateral patella and the lateral malleolus, two muscle strands lateral to the anterior crest of the tibia.
Action : This point is only used reducing to clear phlegm from any location in the body (not just the MBC).

ST.44

Length of needle used : 1"
Depth of insertion : 0.5"
Location : On the dorsum of the foot, at the proximal end of the web between the second and third toes.
Action : This point is only used reducing, it is used to treat three different areas:

1. Support ST.36 when it is done reducing for pain, inflammation, swelling of the MBC relating to the Stomach, Spleen and digestive system in cases of food poisoning and EPA attacking the MBC, gastritis, gastroenteritis, Stomach flu
2. Throat Infections and sore throat, combine with LU.7
3. Trigeminal neuralgia (facial pain), mouth deviation and toothache and jaw problems, combine with LI.4

ST.45

Length of needle used : 0.5"
Depth of insertion : 0.2"
Location : On the lateral side of the second toe, 0.1" from the corner of the nail bed.
Action : Used reducing for arthritis or pain, inflammation or swelling along the second toe. For this moxa can be applied to the handle of the needle. We SB on this point to clear Stomach heat.

6.3.2 Local Extra Points

Lan Wei Xue

Length of needle used : 1.5"
Depth of insertion : 1"
Location : 2" below ST.36.
Action : Used reducing specifically for appendicitis. Apply a very strong reducing manipulation to get benefit.

Zi Gong Xue

Length of needle used : 1.0"
Depth of insertion : 0.5"
Location : Known as the uterus point, it is located 3" beside CV.3.

He Ding

Length of needle used : 1.0"
Depth of insertion : 0.8"
Location : 1" above the patella on the midline.

Xi Yan

Length of needle used : 1.5"
Depth of insertion : 1.2"
Location : Opposite side of the knee to ST.35.

6.3.3 Stomach herbs

Herb	Actions
Lai Fu Zi, Shen Qu, Qing Pi, Ban Xia, Chen Pi, Ji Nei Jin	To clear food / phlegm stagnation in the Stomach and MBC and move downwards
Xuan Shen, Lu Gen, Shi Hu, Mai Men Dong, Zhu Ru, Bai Mao Gen	To clear heat and dryness from the Stomach
Fang Feng, Hai Feng Teng	To clear wind from the Stomach
Bai Zhu, Fu Ling, Da Zao, Gan Cao, Shan Yao	To strengthen Spleen and Stomach and acquired *qi*
Hai Piao Xiao, Gan Cao	To treat Stomach ulcers
Bai Hua She She Cao, Pu Gong Ying, Jin Yin Hua, Da Huang, Gan Cao	For toxins in the Stomach

6.4 The Spleen Meridian

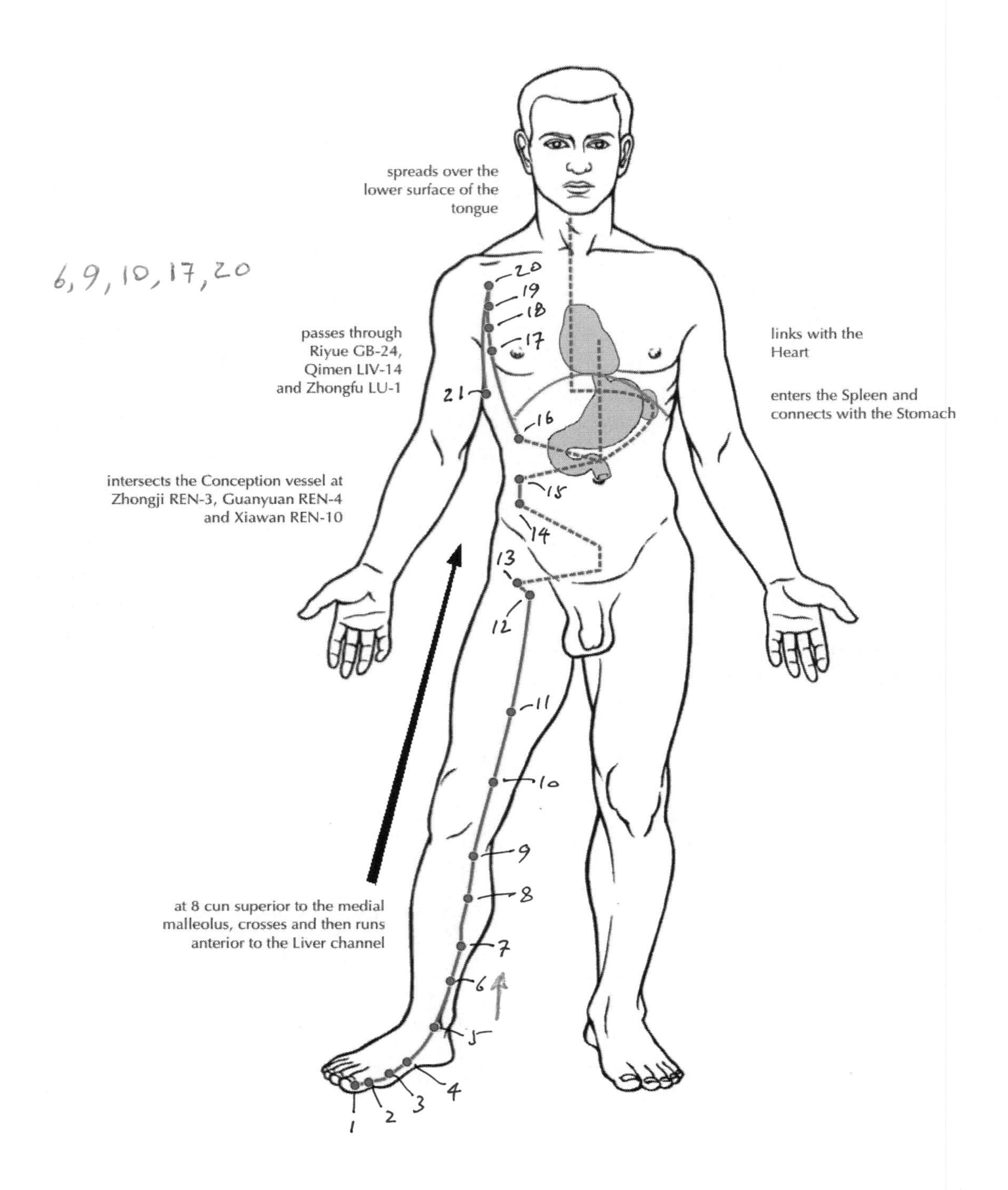

Illustrations reprinted from A Manual of Acupuncture by Peter Deadman & Mazin Al-Khafaji with Kevin Baker by kind permission of the publishers, www.jcm.co.uk.

Discussion

The Spleen is located in the Middle Body Cavity, (MBC). The Spleen makes the blood out of the nutrients it receives from the Stomach.

The Spleen sends nutrient *qi* up to nourish the Lungs and the Heart, the direction of the flow of Spleen *qi* is upwards in the torso and also a strong Spleen will keep the muscles strong and able to lift up. If the Spleen becomes weak then there may be prolapse of the internal organs or diarrhoea. Also if the Spleen *qi* is not nourished by a full and healthy diet then it will not ascend and a person may suffer from dizziness, vertigo or blurred vision. For all these Spleen deficiency conditions we reinforce SP.6 to lift upwards.

The Spleen also nourishes the muscles with the nutrients that it absorbs from food. So if a person is suffering from muscle wasting or weak limbs then to build up the muscles and strengthen the limbs we nourish the Spleen and Stomach with herbs like Gan Cao and Shan Yao and reinforce SP.6 and ST.36. We improve the digestion absorption ability of the Spleen and Stomach with Bai Zhu.

The Spleen also controls the water metabolism and if this ability is impaired there may be diarrhoea, lassitude, abdominal distension, edema (fat and puffy muscle with excess water retention) or problems with phlegm. The Spleen dislikes excessively damp weather.

The Spleen influences the blood vessels because the composition of the blood affects the composition of the blood vessels through which it flows. If a person has a problem with the blood vessels or the blood (easily bruised or easily bleeds), we have to look at the Spleen. The condition of the Spleen and the blood vessels can be observed by looking at the lips. A healthy Spleen results in good appetite and a normal sense of taste.

To quickly replenish the Spleen blood which will help build strong muscles and nourish the Heart and Lungs and boost the immune system we recommend special beef broth to the patient. It is particularly good for for women suffering weaknes due to post menstrual blood loss or post partem (having a baby) exhaustion.

Special beef broth

Beef replenishes, nourishes and tonifies the whole body. It is sweet in flavour and helps the body to make blood. Women should take beef broth post menstrual and post partem to replenish loss of blood, essence and energy so that they can have a strong general health constitution and increased body resistance to disease.

Special Beef Broth is made from shin beef (beef from the shin of the animal), fresh ginger and water. Organic beef is best.

Place one pound of shin beef in a ceramic or glass bowl and fill with water until the beef is covered to a depth of about one inch. Add several slices of fresh ginger and cover. Place this bowl into a pan of water and double boil until the beef goes grey, keeping the outer water topped up.

Dispose of the beef and the ginger. The nutrients from the beef are retained in the water that surrounded it and may be digested more easily rather than having to eat a

Figure 6.1: Double Boiler for Beef Broth recipe

pound of beef. Put the water that surrounded the beef in to a thermos flask so that it stays warm. Drink this broth before and after a meal.

6.4.1 The Spleen points used in clinical practice

SP.1

Length of needle used : 0.5"
Depth of insertion : 0.2"
Location : On the medial great toe, 0.1" from the corner of the nail bed.
Action : Used reducing for arthritis or pain, inflammation or swelling along the big toe, gout. For this moxa can be applied to the handle of the needle. To reduce hyper-menorrhoea (excessive menstrual bleeding or mid cycle bleeding) we SB and then apply 3 Δ to this point in conjunction with SB and apply 3 Δ on LIV.1. This combination would be used after menstruation had finished.

SP.6

Length of needle used : 1.5"
Depth of insertion : 1"
Location : On the medial leg, 3" superior to the medial malleolus, on the posterior border of the tibia. No moxa used on this point.
Action : It is important to understand the many uses of this point, listed here are the most used combinations. It is a commonly used point for a wide variety of conditions.

- **SP.6 Reduced, LI.4 Reinforced** : This combination will create a downwards movement of *qi* in the body so it can be used in combination with LIV.5 reducing prior to menstruation for premenstrual abdominal cramping and pain, endometritis and endometriosis, amenorrhoea (no period). Can also be used to induce child birth if the baby is overdue.
- **SP.6 Reinforced, LI.4 Reduced** : This combination will create a upwards movement of *qi* in the body. Used in combination with LIV.8 reinforcing for deficiency patients who are weak with low immunity, light headed, fainting,

dizziness and vertigo due to blood deficiency in the head. Also for people with organ collapse (visceral prolapse), it could be the Stomach, intestines or anus or sexual organs, we would also add GV.20 reinforcing with 3 Δ.

- **Reduce SP.6, LI.4, ST.25, ST.26, ST.28, ST.36, ST.37, ST.44, GB.34 and LIV.3** : This combination is used reducing for pain, inflammation, swelling of the LBC relating to the digestive system. The abdominal swelling could be caused by wind causing abdominal bloating or food stagnation or food poisoning causing inflammation or EPA attacking the LBC, gastritis, gastroenteritis.
- **Reinforce SP.6 and KID.3, KID.6, ST.36, KID.12, CV.3, LU.1 and KID.27** : to strengthen the Spleen, Liver, Stomach, Lungs and Kidneys this will strengthen the blood, essence and *qi* of the whole body. (to avoid causing an excessive condition we balance these points by reducing LI.4, BL.60, GB.40, GB.34, LU.7, LIV.3 and CV.17). SP.6 has such a strong strengthening effect on the whole body because the kidney and Liver meridians also pass through it. It is known as the three *yin* Meridians Meeting / Crossing Point. Because the the *yin* meridians meet here we can use this point along with KID.3 and BL.23 to replenish the bodies *yin*/essence, to rehydrate the body.

The effects of SP.6, SP.9 and SP.10

If we were to reinforce SP.6 for all the benefits that it offers we could support this by also reinforcing SP.9 and SP.10 as well. However it is important to understand that when reinforcing SP.6 it relates to the LBC, SP.9 to the MBC and SP.10 to the UBC so if we were to reinforce SP.6 and SP.9 and SP.10 we would be lifting the *qi* up through the LBC, MBC and all the way up into the UBC, we would only do this if it was relevant to do so. We do not always add SP.9 and SP.10 when we reinforce SP.6.

So for example if a patient had a weak lower body, it could be weak repoductive ability or weak Kidneys and lower back then SP.6 would be relevant. If they also had weak digestion and prolapse of the internal organs then we could add SP.9 and if they also had a hollow feeling in the chest and Heart with weak respiration and poor circulation light headed, fainting and dizzy then we could add SP.10.

If we were to reduce SP.6 we could support this by also reducing SP.9 and SP.10 as well, however we must remember that the Spleen in TCM is a major component of the immune system and if we are to reduce not only SP.6 but also SP.9 and SP.10 we are going to be having a very strong effect. So either the patient must have a very excessive condition for it to be relevant to reduce all three points or if we need to reduce all three points and we are treating a weak patient then we must reinforce the patient in some other way either by reinforcing KID.12 if they are on their back or BL.20 and BL.23 if they are on their stomachs.

SP.9

Length of needle used : 1.5"
Depth of insertion : 1"

Location : On the medial leg, on the inferior border of the medial condyle of the tibia, in the depression between the posterior border of the tibia and gastrocnemius muscle.
Action : Used reducing for pain, inflammation, swelling of the knee joint, if the problem is on this location. This point is often combined with ST.35 and the extra point on opposite side of knee called Xi Yan.

SP.10 – ***Sea of Blood***

Length of needle used : 1.5"
Depth of insertion : 1"
Location : On the medial thigh, with the knee in flexion, 2" superior to the superomedial angle of the patella, on the vastus medialis muscle.
Action : Used reducing for pain, inflammation, swelling of the knee joint. This point is often combined with ST.35, Xi Yan, and the Extra point 1" below the patella on the midline. Moxa can be applied to the handle of all these needles. SP.10 can also be used as a support point for SP.6 and SP.9, all the uses explained for SP.6 apply to SP.9 and SP.10. Also SP.10 in combination with Bai Chong Wo for heat in the blood and skin rashes.

SP.15

Length of needle used : 1.5"
Depth of insertion : 1"
Location : On the abdomen, at the level of the umbilicus, 4" lateral to the anterior midline.
Action : Used reducing for pain, inflammation, swelling of the MBC and LBC relating to the digestive system. This point is often combined with ST.25 and ST.26 reducing. Moxa can be applied to the handle of the needle.

6.4.2 Local Extra Points

Xi Yan

Length of needle used : 1.5"
Depth of insertion : 1.2"
Location : Eye of the knee, opposite to ST.35.
Action : Knee problems.

He ding

Length of needle used : 1.0"–1.5"
Depth of insertion : 0.8"–1.2"
Location : 1" above the patella on the midline.

Action : Knee problems.

Bai Chong Wo

Length of needle used : 1.0”–1.5”
Depth of insertion : 0.8”–1.2”
Location : 1” proximal to SP.10.
Action : the feature point for heat in the blood and skin rashes, it would be acupunctured in the reducing direction to reduce blood heat, we could also SB to clear heat.

Extra Point

Length of needle used : 1.0”
Depth of insertion : 0.8”
Location 1” below the patella on the midline. Moxa can be applied to the handle of all these needles.
Action : Knee problems.

6.4.3 Spleen herbs

Herb	Actions
Mu Dan Pi, Tao Ren, Dan Shen (Miltiorrhizae)	Clear heat from Spleen
Bai Zhu, Dang Shen (Codonopsis), Gan Cao, Huang Qi, Da Zao,	To strengthen Spleen *qi*
Bai Zhu, Dang Gui, Ji Xue Teng, He Shou Wu	To strengthen Spleen blood
Shan Yao	To strengthen Spleen essence
Shan Yao, Bai zhu, Gan Cao, Dang Gui	To strengthen Spleen to build muscles
Bai Zhu, Fu Ling, Yi Yi Ren, Chen Pi	To strengthen Spleen to clear damp
Shan Yao, Bai Zhu, Chen Pi, Gan Cao, Yi Yi Ren, Sheng Ma, Chai Hu, Huang Qi, Dang Shen (Codonopsis)	To counter Spleen deficiency diarrhoea
Huang Qi, Dang Shen (Codonopsis), Bai Zhu, Gan Cao, Chai Hu, Sheng Ma, Zhi Ke	To lift upwards to counter organ collapse due to weak Spleen

6.5 The Heart Meridian

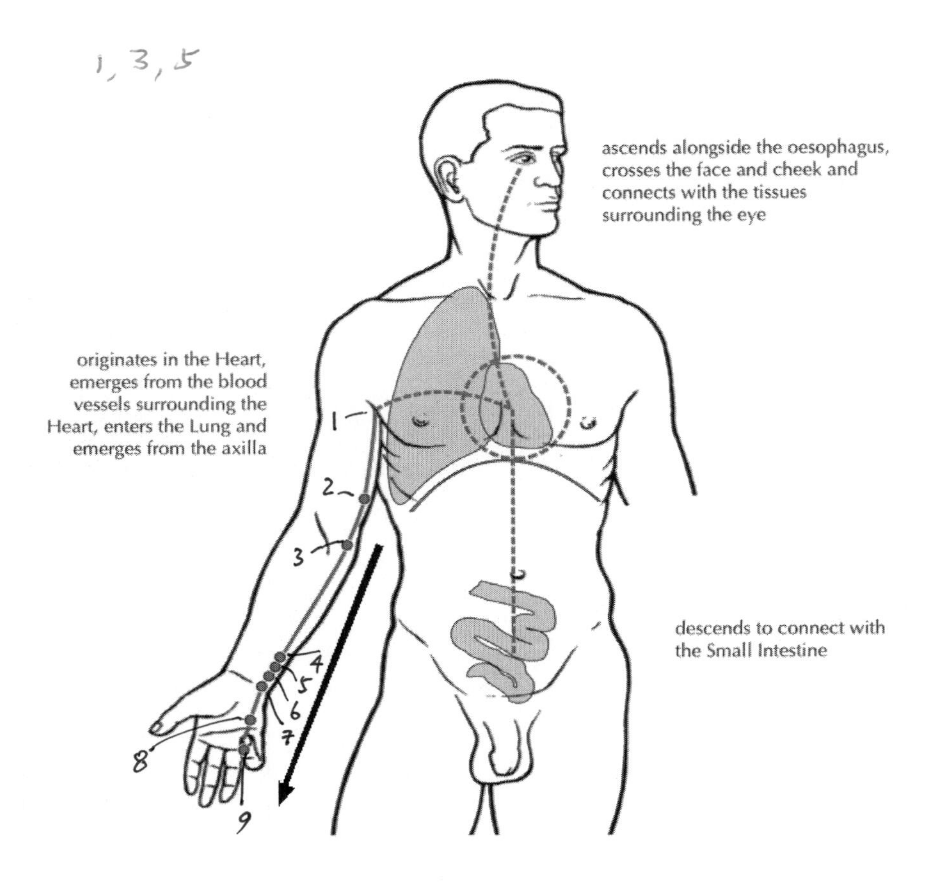

The Heart primary channel

Illustrations reprinted from A Manual of Acupuncture by Peter Deadman & Mazin Al-Khafaji with Kevin Baker by kind permission of the publishers, www.jcm.co.uKID.

Discussion

The Heart is located in the Upper Body Cavity, it pumps blood through the blood vessels to all parts of the body.

An internal branch of the Heart meridian ascends up into the head, so the Heart and the head are connected. So Heart blood deficiency can be the cause of problems with the head such as lightheadedness. For this, we use Chuan Xiong, Dong Kwai, and Ye Jiao Teng. Dang Gui

Also Heart blood deficiency could cause a type of insomnia in which the patient is easily woken due to a loud sound or over dreaming.

The tongue is the opening of the Heart, if there is heat in the Heart the tip of the tongue will be very red, excess Heart heat can cause restlessness and overthinking.

The condition of the Heart can be read from the pulse on the radial artery on the wrist at Stage One left.

The Heart is also reflected in the face, if the face is pale the Heart and circulation is weak.

In summer the excess heat can aggravate the Heart and cause the face to be flushed and the person to be impatient and restless.

6.5.1 The Heart points used in clinical practice

H.3

Length of needle used : 1"
Depth of insertion : 0.5"
Location : With the elbow flexed, at the medial end of the transverse cubital crease.
Action : Used reducing for pain, inflammation, swelling of the elbow joint. Moxa can be applied to the handle of the needle.

H.5

Length of needle used : 1"
Depth of insertion : 0.8"
Location : On the palmar surface of the forearm, 1" proximal to the transverse wrist crease, on the radial side of the flexor carpi ulnaris tendon.
Action : Used reinforcing to strengthen the Heart, the needle is inserted at H.5 and the needle tip reaches almost to H.7. No moxa is used on this point.

H.7

Length of needle used : 0.5"
Depth of insertion : 0.2"
Location : On the transverse wrist crease, in the small depression between the pisiform and ulna bones.
Action : Used reinforcing to strengthen the Heart. Used reducing for pain, heat and inflammation of the Heart and insomnia of the excessive type. Also used for Heart heat causing a person to be restlessness and impatient. This point is commonly used for

calming or relaxing the Heart / mind and spirit in combination with Yin Tang. Used vertical, not reducing, not reinforcing to harmonise the Heart. This method would be used if a patient had acute excessive symptoms of the Heart but constitutional Heart deficiency. No moxa used on this point.

H.9

Length of needle used : 0.5"
Depth of insertion : 0.2"
Location : On the radial side of the fifth digit, 0.1" from the corner of the nail bed.
Action : Used reducing for arthritis or pain, inflammation or swelling along the little finger. For this moxa can be applied to the handle of the needle. We SB to clear acute Heart heat, the tongue tip will be red and the taste buds on the tongue tip will be red and raised.

6.5.2 Local Extra Points

Yin Tang

Length of needle used : 1.0"
Depth of insertion : 0.5"
Location : Between the eyebrows.
Action : This point is translated as *'the gateway to the mind'* and is used reducing to calm the mind. It can also be used reducing for a swollen nose. It would be used reinforcing to boost the local immunity of the nose to strengthen its ability to fight Wind-Heat (hayfever) in summer, and Wind-Cold in winter.

6.5.3 Heart herbs

Herb	Actions
Dang Shen (Codonopsis)	To strengthen poor function of the Heart muscle (Heart miss a beat)
Zi Hua Di Ding, Zhu Ru	To clear Heart heat
Bai He, Gua Lou, (these herbs are balanced with Ban Xia)	To clear Heart heat causing Heart essence deficiency
Suan zao Ren, Bai Zi Ren, Fu Shen, Yuan Zhi, He Huan Pi, Ling Zhi, Long Yan Rou	To calm and sedate the Heart / mind, insomnia and overthinking
Ye Jiao Teng, He Huan Pi	To replenish deficient Heart blood

6.6 The Small Intestine Meridian

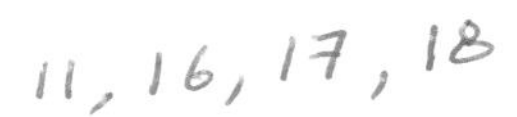

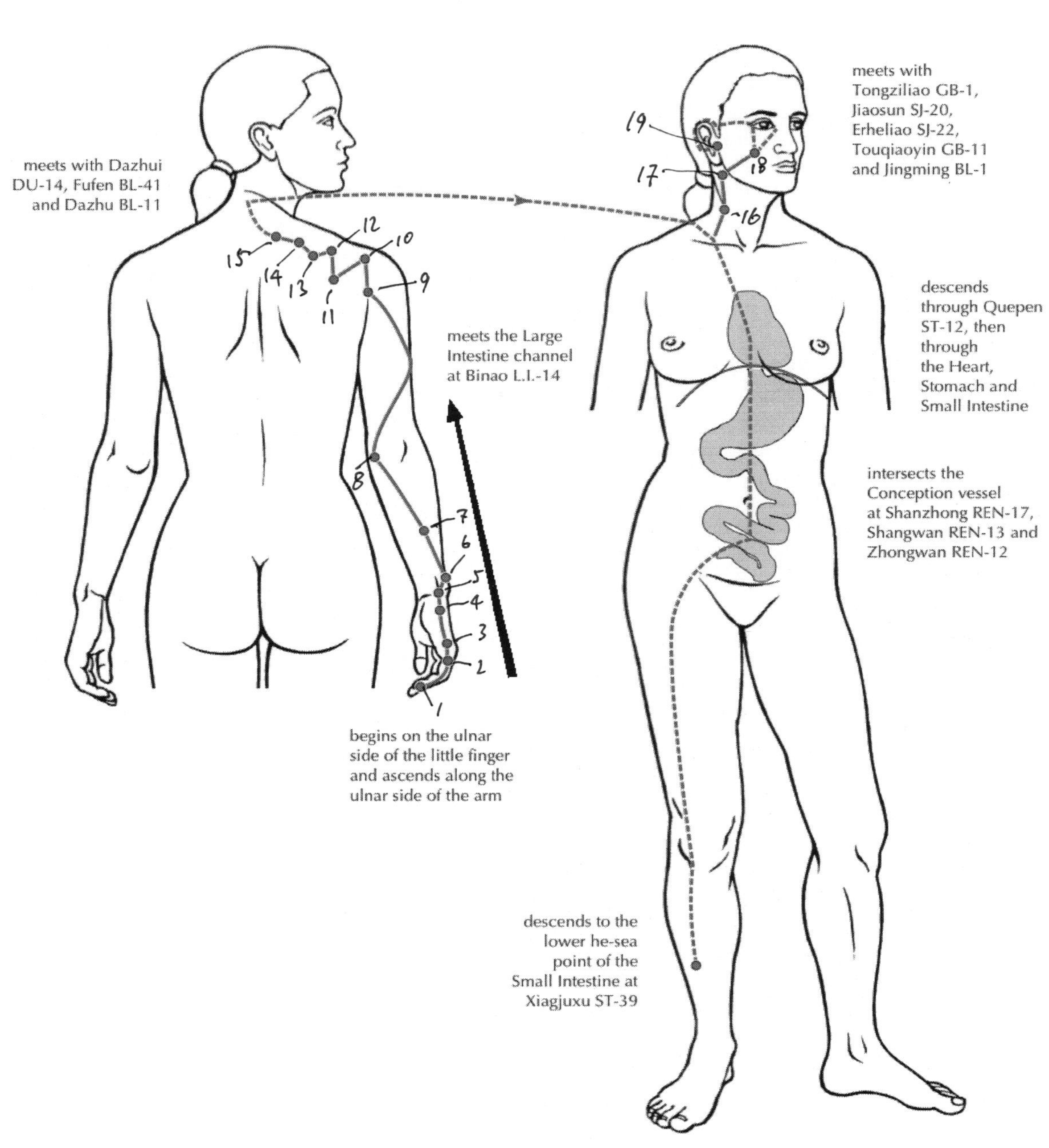

The Small Intestine primary channel

Illustrations reprinted from A Manual of Acupuncture by Peter Deadman & Mazin Al-Khafaji with Kevin Baker by kind permission of the publishers, www.jcm.co.uk.

6.6.1 Discussion

The Small Intestine is located in the Lower and Middle Body Cavity, it receives food from the Stomach at its top end which it further digests and passes on to the Large Intestine at its lower end. Problems with digestion and excretion will affect the Small Intestine.

In clinical practice we mostly use the Small Intestine meridian for upper back and neck problems because the pathway of the meridian covers this area.

6.6.2 The Small Intestine points used in clinical practice

SI.1

Length of needle used : 0.5"
Depth of insertion : 0.2"
Location : On the ulnar side of the fifth digit, 0.1" from the corner of the nail bed.
Action : Used reducing for arthritis or pain, inflammation or swelling along the little finger. For this moxa can be applied to the handle of the needle. We SB to clear acute heat from upper back, neck, eyes and ears.

SI.3

Length of needle used : 1.0"
Depth of insertion : 0.5"
Location : In the depression proximal to the head of the fifth metacarpal bone, at the junction of the dorsal and palmar surfaces. On the apex of the pyramid of skin formed when the hand makes a clenched fist.
Action : Only used reducing to clear pain and inflammation from the upper back, neck and shoulder. This point is used with BL.62, they are paired trigram points. No moxa used on this point.

SI.4

Length of needle used : 1.0"
Depth of insertion : 0.5"
Location : On the ulna side of the wrist, in a depression between the base of the fifth metacarpal and the triquetral bone.
Action : Only used reducing to clear pain and inflammation from the wrist, this point has a vertical insertion due to the local anatomy. Moxa can be applied to the handle of the needle.

SI.5

Length of needle used : 1.0"
Depth of insertion : 0.5"
Location : On the ulnar side of the wrist, in the depression between the ulnar styloid process and the triquetrum and pisiform bones. SI.5 can be located when patient's

wrist is in flexion with the index finger pointing to the sternum.
Action : Only used reducing to clear pain and inflammation from the wrist, this point has a vertical insertion due to the local anatomy. Moxa can be applied to the handle of the needle.

SI.8

Length of needle used : 1.0"
Depth of insertion : 0.5"
Location : With the elbow in flexion, in the depression between the olecranon and the medial epicondyle of the humerus.
Action : Only used reducing for elbow pain. Moxa can be applied to the handle of the needle.

SI.9

Length of needle used : 1.5"
Depth of insertion : 1.0"
Location : On the upper back, with the arm in abduction, 1" superior to the posterior end of the axillary fold.
Action : Only used reducing to clear pain and inflammation from the shoulder, this point has a vertical insertion due to the local anatomy. Moxa can be applied to the handle of the needle.

SI.11

Length of needle used : 1.0"
Depth of insertion : 0.5"
Location : On the scapula, in the depression of the the infrascapular fossa, one-third the distance between the lower border of the scapular spine and the inferior angle of the scapula. Moxa can be applied to the handle of the needle.
Action : Only used reducing to clear pain and inflammation from the shoulder blade (scapula) locally. This point can also be used reducing for breast problems like mastitis (inflammation of the breasts, the feature herb for this is Pu Gong Ying) and breast cysts. Moxa can be applied to the handle of the needle.

SI.12

Length of needle used : 1.0"
Depth of insertion : 0.5"
Location : On the upper back, with the arm in slight abduction, in the middle of the supraspinous fossa.
Action : Only used reducing to clear pain and inflammation and tension from the upper back. Moxa can be applied to the handle of the needle.

SI.13

Length of needle used : 1.0"
Depth of insertion : 0.5"
Location : On the upper back, in the depression on the medial end of the supraspinous fossa.
Action : Only used reducing to clear pain and inflammation and tension from the upper back. Moxa can be applied to the handle of the needle.

SI.14

Length of needle used : 1.0"
Depth of insertion : 0.5"
Location : On the upper back, 3" lateral to the lower border of the spinous process of the first thoracic vertebra (T1).
Action : Only used reducing to clear pain and inflammation and tension from the upper back. Moxa can be applied to the handle of the needle.

SI.15

Length of needle used : 1.0"
Depth of insertion : 0.5"
Location : On the upper back, 2" lateral to the lower border of the spinous process of the seventh cervical verteba (C7).
Action : Only used reducing to clear pain and inflammation and tension from the upper back. Moxa can be applied to the handle of the needle.

6.6.3 Small Intestine herbs

These herbs are not for the Small Intestine organ but for the pathway of the Small Intestine meridian on the upper back. This area is affected by EPA and stress.

Herb	Actions
Mu Gua, Jing Jie, Feng Fang, Qiang Huo	If the upper back and neck is attacked by wind, cold and damp.

6.7 The Bladder meridian

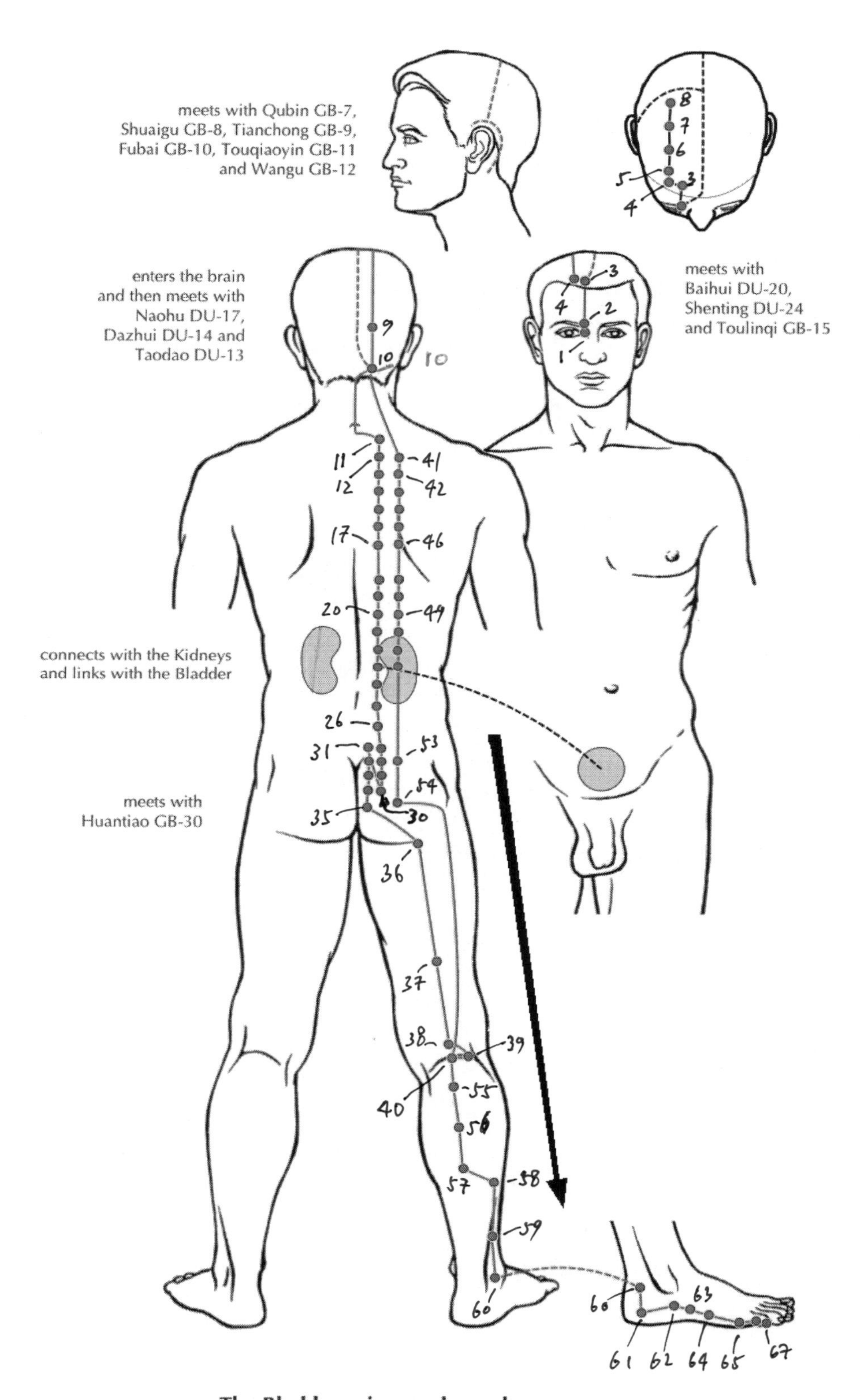

The Bladder primary channel

Illustrations reprinted from A Manual of Acupuncture by Peter Deadman & Mazin Al-Khafaji with Kevin Baker by kind permission of the publishers, www.jcm.co.uk.

Discussion

The Bladder is located in the Lower Body Cavity, it stores and then discharges urine. If the Bladder *qi* is weak it can lead to incontinence of urine. If there is heat in or infection of the Bladder it can cause pain on urination, less urination or darker yellow urine.

The Bladder meridian goes over the head so we can use it for eye problems and head ache. It also runs down the leg so we use these points for sciatica pain which goes down the leg. The meridian also covers much of the back, so we use points on this meridian that are over the internal organs to control the internal organs, 90% of the time we use these Shu points in the reinforcing direction to tonify the internal organs.

When we reinforce the Shu points on the Bladder meridian we often reduce the GV point that is on the same level to balance our treatment. However when we reinforce BL.20 Spleen Shu we do not do the GV because we do not want to reduce the Spleen *yang*.

There is a medical condition called pneumothorax which means a puncture of the thoracic cavity, in seriouse cases this results in air entering in to the thoracic cavity and the collapse of the lung. If pneumothorax was caused by an acupuncture needle and the patient went to hospital they would be put under observation and no action would be taken because the hole was so small and the body would heal itself.

It is only when there is a major traumatic injury that a procedure is used. Nontheless to avoid the occurance of pneumothorax we always insert the needles at an angle of about 25 degrees when acupuncturing on the back above thoracic vertebra 12 (T 12) on all points on the Bladder meridian and the Small Intestine meridian.

When acupuncturing on the front of the torso it is the same, so when the most used points, L.1 and KID.27 are acupunctured the needles are inserted at an angle of about 25 degrees. However for CV.17 the needle is almost flat against the skin, this is for two reasons, first the sternum (breastbone) is just under this point and also because there is a rare condition called sternal foramen, (a hole in the sternum) and to avoid going through this in to the thoracic cavity and causing pneumothorax we have a very flat angle of insertion for the acupuncture needle.

6.7.1 The Bladder points used in clinical practice

BL.2

Length of needle used : 1.5"
Depth of insertion : 0.5"
Location : At the medial end of the eyebrow, on the supraorbital notch.
Action : Only used reducing for pain, inflammation, swelling and involuntary spasm of the area around the eye. For anatomical reasons this point is always acupunctured in the same direction, towards the extra point Yu Yao located at the peak of the eyebrow. For BL.2 a long needle is required as we apply moxa and need safe distance from face. This point is often combined with GB.14 which is also punched in the direction of Yu yao.

BL.3

Length of needle used : 1.0"
Depth of insertion : 0.5"
Location : In the frontal region of the head, 0.5" within the anterior hairline and 0.5" lateral to GV.24 at the midline.
Action : Only used reducing for pain, inflammation, swelling and congestion of the nose and frontal headache. Moxa can be applied to the handle of the needle. For nose problems we often support our use of BL.3 with LI.20 and the three extra points, Bi Tong (above LI.20), Yin Tang (third eye point between eyebrows) and the high point on the bridge of the nose. These points are all local, the distant points to support this work would be LU.7 reducing, this is because the nose is the opening of the Lungs. Also LI.4 reducing because it controls the head and face.

BL.7

Length of needle used : 1.0"
Depth of insertion : 0.5"
Location : On the head, 4" posterior to anterior hairline, 1.5" lateral to the midline.
Action : Only used reducing for pain, inflammation and swelling and congestion of the nose. Used to reduce nasal congestion, anosmia (no smell) and epistaxis (nose bleed). Moxa can be applied to the handle of the needle.

BL.10

Length of needle used : 1.0"
Depth of insertion : 0.5"
Location : 0.5" inside the hair margin on the lateral side of the tendon. This point is inserted in the direction of the other BL.10 due to the local anatomy.
Action : Only used reducing for pain, swelling and inflammation in the local occipital area and neck. Used to reduce nasal congestion, anosmia (no smell) and epistaxis (nose bleed). Also used reducing to clear the whole head in conjunction with GB.20 (which is punched towards the opposite eye) for headaches, migraine and wind, cold and damp attack to the head. Moxa can be applied to the handle of the needle.

BL.11

Length of needle used : 1.0"
Depth of insertion : 0.5"
Location : On the upper back, 1.5" lateral to the lower border of the spinous process of the first thoracic vertebra (T1).
Action : Only used reducing to clear Exogenous Pathogenic Wind Attack on the Lungs. This point has a shallow angle of insertion, about 25 degrees to avoid pneumothorax (puncture of the thoracic cavity). For wind EPA this point is often combined with

BL.12 reducing, both points are covered with a Cup after the needle has been removed. Also GB.14 can reduce and clear wind attacks to upper back when acupunctured reducing. (There is a special use for BL.11 reinforcing; this is when the patient has severe degeneration of the bones in the whole body such as osteoporosis. In this case we combine BL.11 with BL.23, BL.24, BL.52, KID.3 and KID.6).

BL.12

Length of needle used : 1.0"
Depth of insertion : 0.5"
Location : On the upper back, 1.5" lateral to the lower border of the spinous process of the second thoracic vertebra (T2).
Action : Only used reducing to clear Exogenous Pathogenic Wind Attack on the Lungs. This point has a shallow angle of insertion, about 30 degrees to avoid pneumothorax (puncture of the thoracic cavity). For wind EPA this point is often combined with BL.11, both points are covered with a cup after the needle has been removed. Also GB.14 can reduce and clear wind attacks to upper back when acupunctured reducing.

BL.13 – Lung Shu

Length of needle used : 1.0"
Depth of insertion : 0.5"
Location : On the upper back, 1.5" lateral to the lower border of the spinous process of the third thoracic vertebra (T3).
Action : Only used reinforcing to strengthen weak Lungs for chronic and acute lung problems. Moxa can be applied to the handle of the needle. 3 Δ.

BL.14 – Pericardium Shu

Length of needle used : 1.0"
Depth of insertion : 0.5"
Location : On the upper back, 1.5" lateral to the lower border of the spinous process of the fourth thoracic vertebra (T4).
Action : Only used reinforcing to strengthen weak Lungs for chronic and acute lung problems, this point is used to support Lung Shu. Moxa can be applied to the handle of the needle. 3 Δ.

BL.15 – Heart Shu

Length of needle used : 1.0"
Depth of insertion : 0.5"
Location : On the upper back, 1.5" lateral to the lower border of the spinous process

of the fifth thoracic vertebra (T5).
Action : Only used reinforcing to strengthen weak Heart for chronic and acute Heart problems. Moxa can be applied to the handle of the needle. 3 Δ.

BL.16 – Governing Shu

Length of needle used : 1.0"
Depth of insertion : 0.5"
Location : On the upper back, 1.5" lateral to the lower border of the spinous process of the sixth thoracic vertebra (T6).
Action : Only used reinforcing to strengthen weak Heart for chronic and acute Heart problems, this point is used to support Heart Shu. Moxa can be applied to the handle of the needle. 3 Δ.

BL.17 – Diaphragm Shu

Length of needle used : 1.0"
Depth of insertion : 0.5"
Location : On the middle back, 1.5" lateral to the lower border of the spinous process of the seventh thoracic vertebra (T7).
Action : Only used reinforcing to strengthen the diaphragm and to replenish the blood, It is the Influential Point of the Blood, being located between the Heart point and the Liver point. The Liver stores blood while the Heart moves the blood. This point is situated between the two organs and has the function to promote the manufacture of blood. It is used for treating disorders of blood such as anemia. If we reinforce this point combined with BL.18, BL.20, LIV.8, SP.6 and KID.3 it will replenish the Liver blood. Moxa can be applied to the handle of the needle. 3 Δ.

BL.18 – Liver Shu

Length of needle used : 1.0"
Depth of insertion : 0.5"
Location : On the middle back, 1.5" lateral to the lower border of the spinous process of the ninth thoracic vertebra (T9).
Action : Only used reinforcing to strengthen a weak Liver that is recovering from a chronic Liver problem like hepatitus. If we reinforce this point combined with BL.17, BL.20, LIV.8, SP.6 and KID.3 it will replenish the Liver blood. No moxa used on this point and no moxa cones to avoid creating Liver Heat.

BL.20 – Spleen Shu

Length of needle used : 1.0"
Depth of insertion : 0.5"

Location : On the middle back, 1.5" lateral to the lower border of the spinous process of the eleventh thoracic vertebra (T11).
Action : Only used reinforcing to strengthen a weak Spleen and to boost the immune system. When we reinforce the Shu points on the Bladder meridian we often reduce the GV point that is on the same level to balance our treatment. However when we reinforce BL.20 Spleen Shu we do not do the GV because we do not want to reduce the Spleen *yang*. Moxa can be applied to the handle of the needle. 3 Δ.

BL.23 – Kidney Shu

Length of needle used : 1.0", 1.5" for a large person.
Depth of insertion : 0.5", or 1.0" for a large person.
Location : On the lower back, 1.5" lateral to the lower border of the spinous process of the second lumbar vertebra (L2).
Action : Only used reinforcing to strengthen weak Kidneys and weak kidney/sexual function, impotence/infertility, weak lower back and to boost the immune system and strengthen the bones. This point is often combined with BL.24 and BL.52 to further strengthen the reinforcing action on the Kidneys. Moxa can be applied to the handle of the needle. 3 Δ.

BL.24

Length of needle used : 1.0", 1.5" for a large person.
Depth of insertion : 0.5", or 1.0" for a large person.
Location : On the lower back, 1.5" lateral to the lower border of the spinous process of the third lumbar vertebra (L3).
Action : Only used reinforcing to strengthen weak Kidneys and weak kidney/sexual function, impotence/infertility, weak lower back and to boost the immune system and strengthen the bones. This point is used to support Bl 23 which is further supported by BL 52 to strengthen the reinforcing action on the Kidneys. Moxa can be applied to the handle of the needle. Δ x 3.

6.7.2 Bladder meridian points on the back 3" beside the midline

BL.43

Length of needle used : 1.0"
Depth of insertion : 0.5"
Location : On the upper back, 3" lateral to the lower border of the spinous process of the fourth thoracic vertebra (T4), at the level of BL 14.
Action : 3 Δ are often used on this point to strengthen weak Lungs and to boost the immune system by strengthening and nourishing the lung, Heart, Kidneys, Spleen and Stomach. It is used for all lung deficiency conditions and for all symptoms of poor general health, asthma, weak chronic cough, perspiration at night, weak respiration,

poor digestion, and weak kidney/sexual function, impotence/infertility. This point is only acupunctured in the reinforcing direction to tonify deficiency conditions.

BL.52

Length of needle used : 1.0”, 1.5” for a large person.
Depth of insertion : 0.5”, or 1.0” for a large person.
Location : On the lower back, 3” lateral to the lower border of the spinous process of the second lumbar vertebra (L2), at the level of BL 23.
Action : Only used reinforcing to strengthen weak Kidneys and weak kidney/sexual function, impotence/infertility, weak lower back and to boost the immune system and strengthen the bones. This point is used to support Bl 23 which is further supported by BL 24 to strengthen the reinforcing action on the Kidneys. Moxa can be applied to the handle of the needle. 3 Δ.

6.7.3 The Four Sacral Points

BL.31

Length of needle used : 2.5”
Depth of insertion : 2.0”
Location : In the sacral region, in the first posterior sacral foramen. Moxa can be applied to the handle of the needle.

BL.32

Length of needle used : 2.5”
Depth of insertion : 2.0”
Location : In the sacral region, in the second posterior sacral foramen. Moxa can be applied to the handle of the needle.

BL.33

Length of needle used : 2.5”
Depth of insertion : 2.0”
Location : In the sacral region, in the third posterior sacral foramen. Moxa can be applied to the handle of the needle.

BL.34

Length of needle used : 2.5”
Depth of insertion : 2.0”

Location : In the sacral region, in the fourth posterior sacral foramen. Moxa can be applied to the handle of the needle.

The Action of the four sacral points BL.31, BL.32, BL.33, BL.34

When the patient has a deficiency condition of the sexual organs such as impotence or infertility or if the patient is suffering from insufficient blood and essence we can reinforce BL.31, BL.32, BL.33, BL.34.

If the patient has an excessive condition of the genital urinary system with pain swelling or inflammation, such as an infection, either viral or bacterial, or if they have blood stagnation, piles, or haemorroids then we do not reduce BL.31, BL.32, BL.33, or BL.34. Instead we reduce Hua Tuo and GV points on the sacrum.

BL.36

Length of needle used : 1.5"
Depth of insertion : 1.0"
Location : In the gluteal region, in the middle of the transverse gluteal fold.
Action : Only used reducing for sciatica pain running down the back of the leg. Moxa can be applied to the handle of the needle.

BL.37

Length of needle used : 1.5"
Depth of insertion : 1.0"
Location : On the posterior thigh, 6" inferior to BL 36.
Action : Only used reducing for sciatica pain running down the back of the leg. Moxa can be applied to the handle of the needle.

BL.40

Length of needle used : 1.5"
Depth of insertion : 0.5"-1.0"
Location : At the midpoint of the popliteal fossa.
Action : This point has a vertical insertion due to the local anatomy, it is only used reducing (achieved through needle manipulation) to clear the back of the torso prior to working on the back of the torso and to release the back of the knee if it is tight or painful, we also use this point if the sciatica pain running down the back of the leg reaches this far. Moxa can be applied to the handle of the needle.

BL.57

Length of needle used : 1.0"
Depth of insertion : 0.5"
Location : On the posterior leg, 8" inferior to BL 40.
Action : Only used reducing for tightness or spasm of the calf muscle, haemorrhoids and constipation. Moxa can be applied to the handle of the needle.

BL.59

Length of needle used : 1.0"
Depth of insertion : 0.5"
Location : On the posterior surface of the leg, 3" superior to BL.60 posterior to the lateral malleolus.
Action : Only used reducing for lower back pain and sciatica pain running down the back of the leg. Moxa can be applied to the handle of the needle.

BL.60

Length of needle used : 1.0"
Depth of insertion : 0.5"
Location : On the lateral ankle, in the depression midway between the external malleolus and the tendon calcaneus.
Action : Only used reducing to clear headache, lower back pain and sciatica pain running down the back of the leg. This point also clears heat from the Bladder so it can be used if there is inflammation and infection in the Bladder or kidney. We also do this point reducing to balance KID.3 reinforcing.

BL.62

Length of needle used : 1.0"
Depth of insertion : 0.5"
Location : On the lateral foot, in the depression directly inferior to the lateral malleolus, at the dorsal-plantar skin junction.
Action : Only used reducing to clear pain and inflammation from the upper back, neck and shoulder. This point is used with SI.3 reducing, they are paired trigram points.

6.7.4 Bladder herbs

Herb	Actions
Fu Pen Zi, Jin Ying Zi, Wu Wei Zi, Lian Xu, Shan Zhu Yu	If there is a weak Bladder causing excessive urination.
Hua Shi, Che Qian Cao, Bai Mao Gen, Chi Shao Yao, Lu Gen, Huai Hua Mi, Pu Gong Ying, Jin Yin Hua, Lian Qiao, Tu Fu Ling	If there is an infection in the urinary tract and Bladder.

6.8 The Kidney Meridian

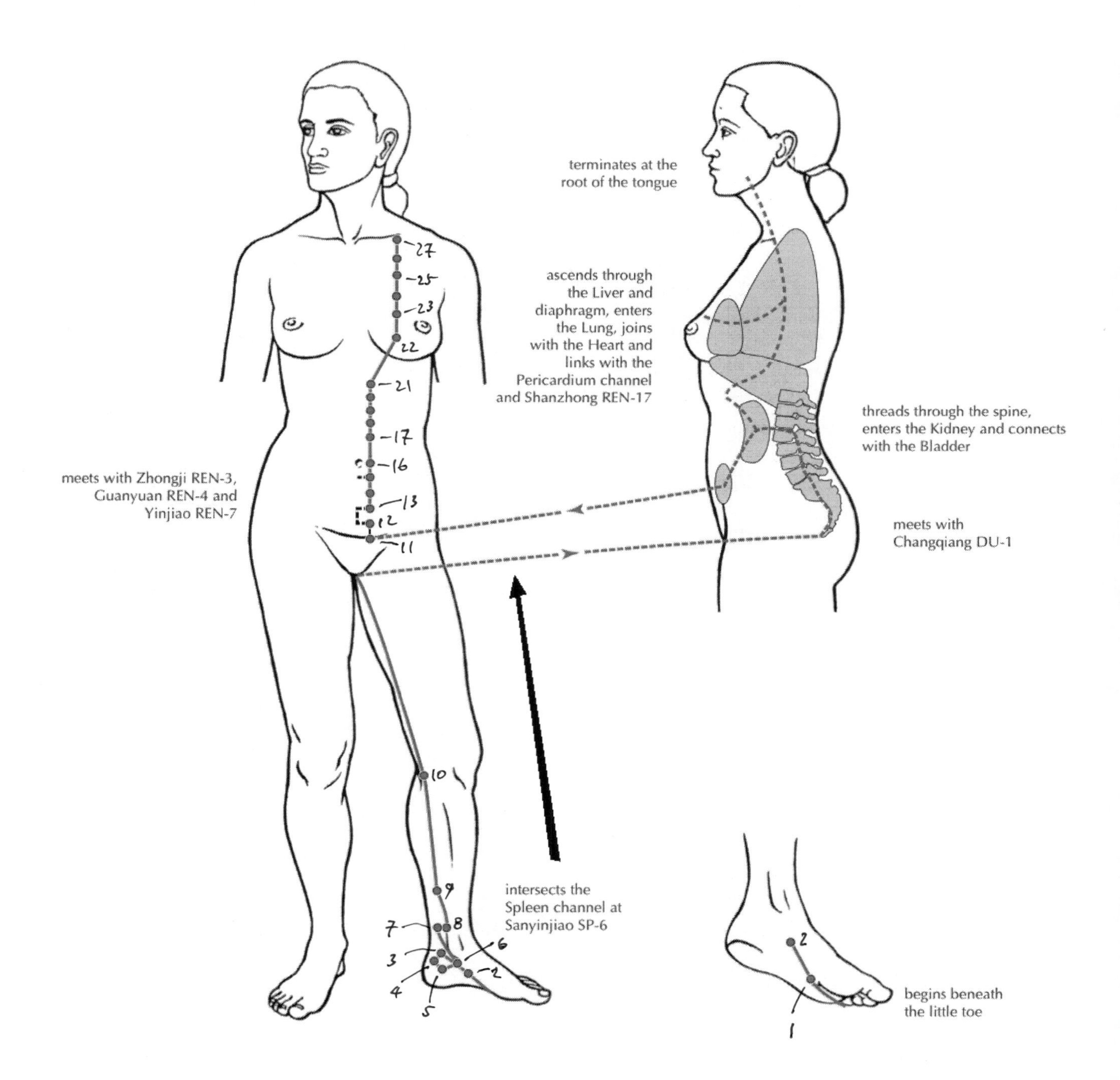

Illustrations reprinted from A Manual of Acupuncture by Peter Deadman & Mazin Al-Khafaji with Kevin Baker by kind permission of the publishers, www.jcm.co.uk.

Discussion

The Kidneys are located in the Lower Body Cavity, on either side of the lumbar vertebrae. They create and store essence and are associated with inherited energy. They control the Lower Body Cavity and so influence reproduction and the sexual organs and the Bladder and urination. Because the Kidneys are involved with regulating water in the body we look to it in cases of edema (as well as to the Spleen and lung). Because the kidney meridian runs up the legs if the Kidneys become weak the legs can also become weak, particularly the knees.

The Kidneys also control the composition of the bones (particualy the lumbar vertebrae) so as part of an anti arthritis treatment we would strengthen the Kidneys. The Kidneys also control the teeth which are the ends of the bones, if the Kidneys are weak the teeth will easily chip and be loose.

If the hearing is becoming weak in old age it is often due to weakening of the Kidneys because the Kidneys control the ears. When treating tinitus (ringing in the ears) we also treat the Kidneys. The Kidneys are also filters so they may ache when a person has caught an Exogenous Pathogenic Attack as they are trying to filter it.

Overwork and illness can consume essence and weaken the Kidneys causing them to ache, also with men excess loss of seminal essence will cause the Kidneys to ache and become weaker. Kidney essence nourishes the bone marrow, spinal fluid and brain, hormones and blood. Severe depletion of the Kidneys will cause the lower back to feel sore around and between the lumbar vertebrae, particularly below lumbar 4 and 5. In these situations people are more likely to injure their lower back when lifting something due to this pre-existing weakness.

The Kidneys also influence the hair, as a person gets older and their Kidneys get weaker, their hair goes grey and falls out (To strengthen the hair nourish the Liver blood and kidney essence with He shou Wu and Nu Zhen Zi). Black is the colour of the Kidneys and people with health problems which weaken the Kidneys will have black rings under the eyes.

The taste associated with the Kidneys is salty, consumption of too much salt will damage them and could lead to high blood pressure because the Kidneys control the Heart.

In ancient Chinese Traditional Philosophy the Kidneys are associated with water and the Heart with fire and it is said that the water and fire regulate each other. In modern words we would say that the adrenalin secreted by the adrenal glands which are on top of the Kidneys regulate the Heart function. So to replenish kidney essence deficiency and calm the Heart we reinforce BL.23, BL.24 and BL.52 and balance this with GV.4 reducing. And to clear excess Heart fire we reduce GV below T5 and T6 and Hua Tuos points below T5 and T6. This is also a useful formula to reduce stress and balance spiritual and emotional problems.

The Kidneys dislike the cold and people with bone injuries often complain that when it gets cold in winter their injuries feel worse. Patients should always keep their Kidneys covered and warm in cold weather.

Every organ in the body has its own Essence/Hormones/Water/*yin* and its own *qi*/functional activity/Fire/*yang*. However it is the Essence/Hormones/Water/*yin* and the *qi*/functional activity/Fire/*yang* of the Kidneys that they are all dependent upon.

The right Kidney is *yin*, the left Kidney is *yang*.

If the Kidneys decline the whole body will weaken, if the Kidneys are kept strong the whole body will be strong. We nourish our Kidneys by reinforcing, KID.3, KID.6, KID.7, KID.12, CV.4, BL.23, BL.24 and BL.52.

6.8.1 The Kidney points used in clinical practice

KID.3

Length of needle used : 1.0"
Depth of insertion : 0.5"
Location : On the medial ankle, at the midpoint in the depression between the prominence of the medial malleolus and achille's tendon.
Action : Only used reinforcing to strengthen the Kidneys in general and more specifically the kidney essence and the essence of the whole body.

KID.6

Length of needle used : 1.0"
Depth of insertion : 0.5"
Location : On the medial ankle, in the depression 1" inferior to the medial malleolus.
Action : Only used reinforcing to strengthen the Kidneys in general and more specifically to help the Kidneys support the Lungs. We often use KID.6 reinforcing with LU.7 reducing, they are paired trigram points. So if we were using LU.7 reducing to clear a sore throat, headache, EPA we would support this with KID.6 reinforcing.

KID.7

Length of needle used : 1.0"
Depth of insertion : 0.5"
Location : On the medial leg, 2" superior to KID.3 on the anterior border of the Achilles tendon.
Action : Only used reinforcing to strengthen the Kidneys in general and more specifically kidney *yang*. This is one of the Additional Paired points, KID.7 reinforcing combined with LI.4 reducing is used to stop sweating.

KID.12

Length of needle used : 1.0"
Depth of insertion : 0.5"
Location : In the pubic region, 1" superior to the symphysis pubis, 0.5" lateral to the anterior midline, and at the level of CV.3 which is 4" below the navel.
Action : Only used reinforcing to strengthen the Kidneys *yin* and *yang*. Moxa can be applied to the handle of the needle. It has similar uses to BL.23. To strengthen weak

Kidneys and weak kidney/sexual function, impotence/infertility, weak lower back and to boost the immune system. If we wanted to reinforce BL.23 but could not reach it because the patient was having a treatment whilst lying on their back we could use KID.12 instead.

KID.19

Length of needle used : 1.0"
Depth of insertion : 0.5"
Location : In the epigastric region, 4" superior to the umbilicus and 0.5" lateral to the midline at the level of CV.12.
Action : Only used reducing to clear pain, swelling, inflammation from the MBC. Moxa can be applied to the handle of the needle. KID.19 can reduce food stagnation and is used to support CV.12 reducing.

KID.27

Length of needle used : 1.0"
Depth of insertion : 0.5"
Location : On the lower border of the clavicle, 2" lateral to the anterior midline.
Action : Only used reinforcing to strengthens the Lungs and can also strengthen the Heart, this point has a shallow angle of insertion, about 25 degrees to avoid pneumothorax (puncture of the thoracic cavity). Moxa sticks can be applied to the handle of the needle and birds pecking technique can be used to strengthen weak Lungs and for more severe deficiency cases 3 Δ can be used directly on this acupuncture point after the needle has been removed. This point is often combined with LU.1 reinforcing.

6.8.2 Kidney herbs

Herb	**Actions**
Nu Zhen Zi, Gou Qi Zi, Han Lian Cao, Mai Men Dong, Tian Men Dong, He Shao Wu	For deficiency of kidney *yin*/essence
Du Zhong, Xian Mao, Tu Si Zi, Yin Yang Huo,	For deficiency of kidney *yang*/functional power

6.9 The Pericardium Meridian

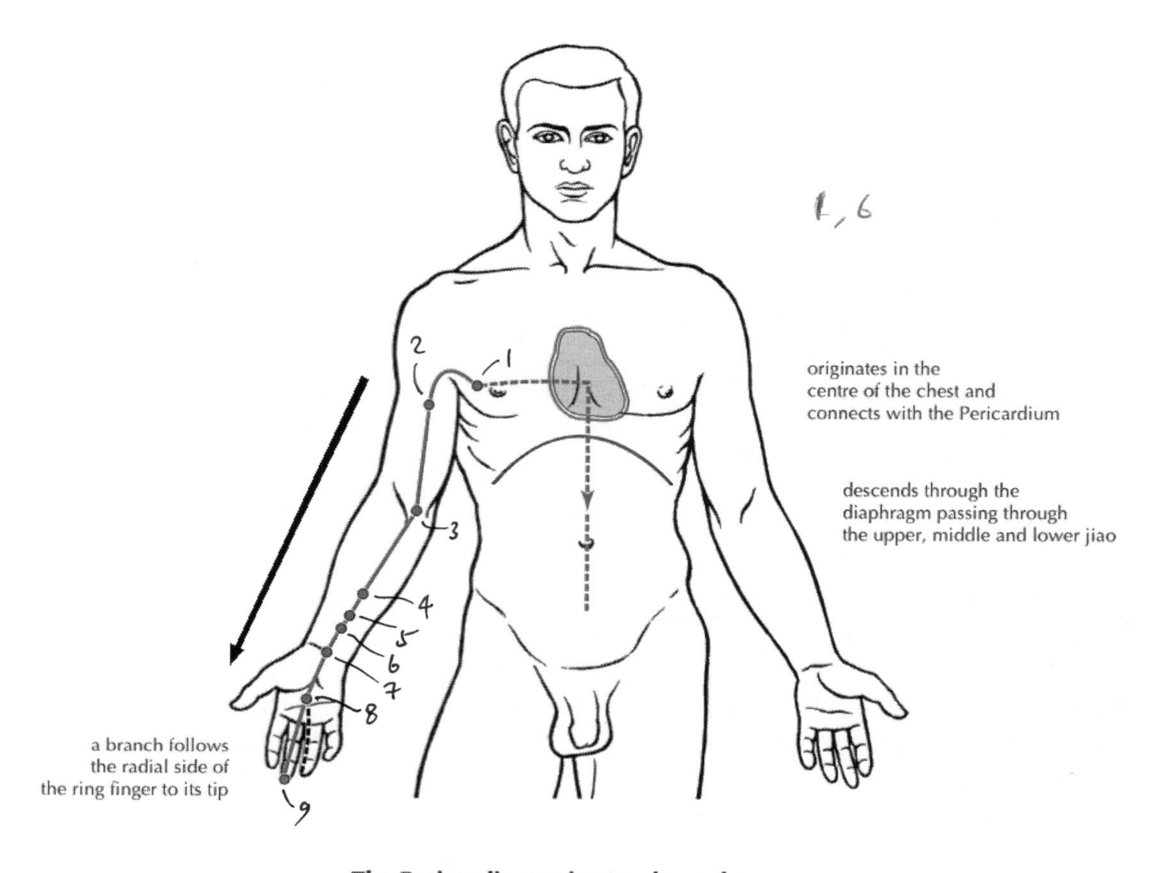

The Pericardium primary channel

Discussion

The function of the Pericardium is to protect the Heart from Exogenous Pathogenic Factors. The Pericardium meridian, as well as going through the Heart and Pericardium, also goes through the digestive tract. This is why P.6 can be used for nausea.

6.9.1 The Pericardium points used in clinical practice

P.6

Length of needle used : 1.0"
Depth of insertion : 0.5"
Location : On the anterior forearm, 2" superior to the transverse wrist crease, between the tendons of palmaris longus and flexor carpi radialis.
Action : Only used reducing for digestive problems like nausea and vomiting. Also used reducing to clear heat, pain, inflammation and stress from the Heart and Pericardium and chest. P.6 reducing and SP.4 reinforcing are paired trigram points used for treating diabetes.

P.9

Location : On the centrE of the tip of the middle finger.
Action : We SB on this point to clear heat from the Pericardium and Heart, most often this will occur in summer.

6.10 Triple Warmer Meridian

5, 10, 11, 12, 17

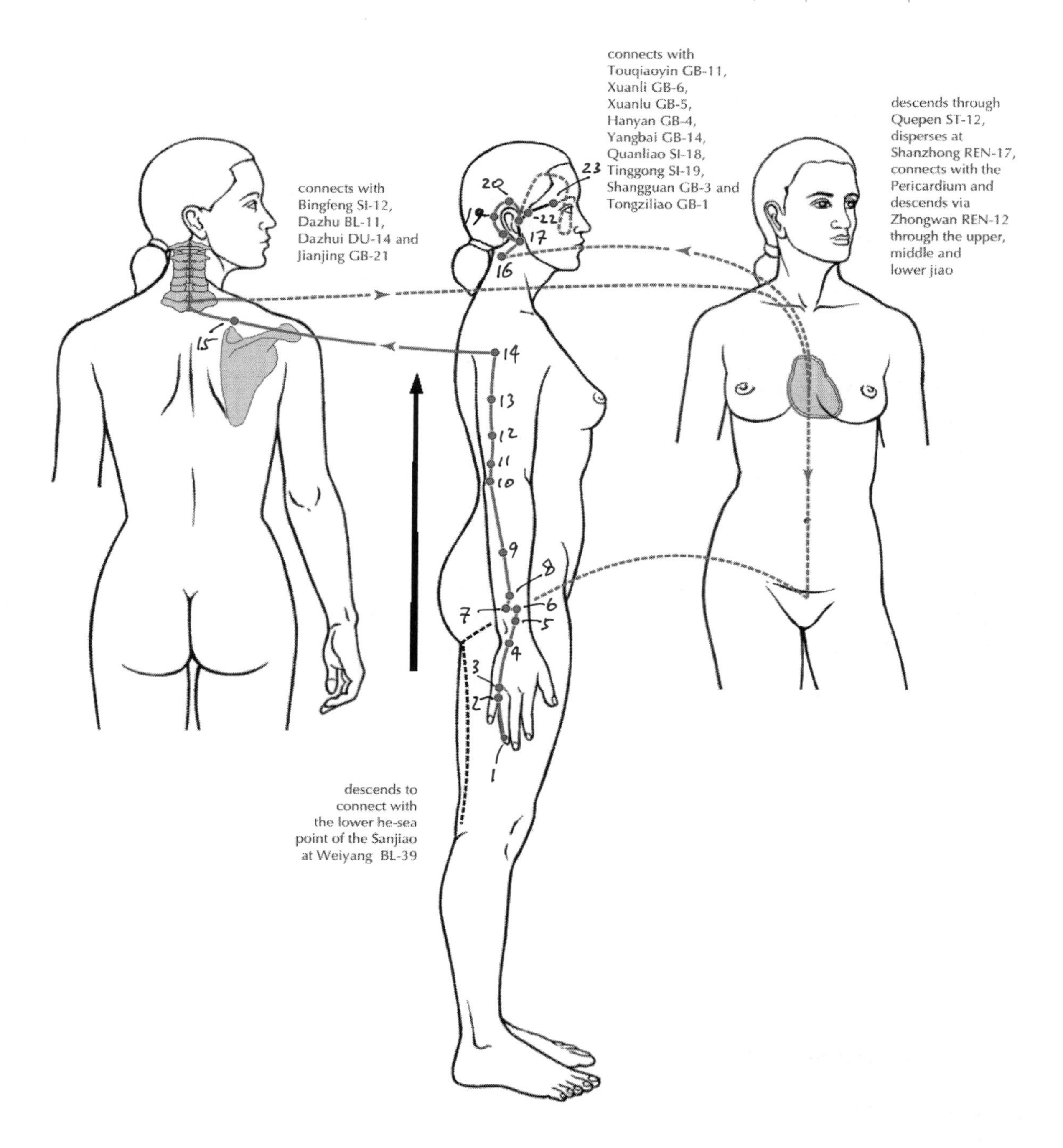

The Sanjiao primary channel

Illustrations reprinted from A Manual of Acupuncture by Peter Deadman & Mazin Al-Khafaji with Kevin Baker by kind permission of the publishers, www.jcm.co.uk.

Discussion

The three warmers are, the Upper Body Cavity (above the solar plexus), the Middle Body Cavity (between the solar plexus and the navel) and the Lower Body Cavity (below the navel).

This meridian runs through all three body cavities and helps to regulate the temperature between these three areas of the body.

The UBC is the heat of respiration and circulation, the MBC is the heat of digestion and the LBC is the heat of reproduction.

The Triple Warmer also connects with the Pericardium. The Triple Warmer meridian also helps to distribute the essence of the Kidneys around the body. It also helps to distribute body fluids. It is primarily used to disperse the build up of excess heat / temperature.

6.10.1 The Triple Warmer Points used in clinical practice

TW.3

Length of needle used : 1.0"
Depth of insertion : 0.5"
Location : On the dorsum of the hand, in the depression proximal to the fourth and fifth metacarpophalangeal joints.
Action : Used reducing for pain, inflammation, swelling of the ears. This is the distant point for the ears, it would be used in combination with the local points GB.2 and TW.21. TW.3 is also the location for the extra point called Luo Zhen which is used reducing for pain, inflammation, swelling of the upper back and neck.

TW.4

Length of needle used : 1.5"
Depth of insertion : 1.0"
Location : On the dorsal transverse wrist crease, between the tendons of muscles extensor digitorum and extensor digiti minimi.
Action : Used reducing for pain, swelling and inflammation of the wrist.

TW.5

Length of needle used : 1.0"
Depth of insertion : 0.5"
Location : On the dorsal forearm, between the radius and ulna, 2" superior to TW.4 at the dorsal transverse wrist crease.
Action : Only used reducing to clear heat / temperature from the whole body (febrile diseases, temperature, low grade fever and heat causing a rash, overall this point works more on the UBC). This point is often combined with GB.41, they are paired trigram

points both used reducing to clear heat from the whole body.

TW.10

Length of needle used : 1.0"
Depth of insertion : 0.5"
Location : In the posterior cubital region, in the depression 1" superior to the olecranon when the elbow is flexed.
Action : Used reducing for pain, swelling and inflammation of the elbow.

TW.13

Length of needle used : 1.5"
Depth of insertion : 1.0"
Location : Two thirds of the distance between TW.10 and TW.14, on the posterior border of the deltoid muscle.
Action : Used reducing for pain, swelling and inflammation of the the shoulder, this is the adjacent point used to support TW.14.

TW.14

Length of needle used : 2.5"
Depth of insertion : 2.0"
Location : At the origin of the deltoid muscle, in the depression which lies posterior and inferior to the lateral extremity of the acromion.
Action : Used reducing for pain, swelling and inflammation of the shoulder, this is the local point. This point is always inserted pointing down the arm towards the elbow due to the local anatomy, this direction is the reducing direction. If we wanted to reinforce the point because the patient had a weak shoulder due to post stroke or muscle wasting then we would get a reinforcing effect through the acupuncture autoscope machine. Moxa can be applied to the handle of the needle for both excessive and deficient shoulder problems.

TW.21

Length of needle used : 1.5"
Depth of insertion : 1.0"
Location : Anterior to the ear, in the depression anterior to the supratragic notch and posterior to the mandibular condyloid process when the mouth is open.
Action : Used reducing for pain, swelling, congestion and inflammation of the ear. Used reinforcing for loss of hearing. This point has a vertical insertion due to the local anatomy. We would get a reducing or reinforcing effect through our needle

manipulation, the acupuncture autoscope machine and through moxa applied to the handle of the needle. This point is often combined with GB.2 and TW.3

How to insert and remove TW.21: We ask the patient to open their mouth wide, we then insert the needle, they then close their mouth. To remove the needle we ask the patient to open their mouth wide, we then remove the needle, they then close their mouth.

6.10.2 Triple Warmer herbs

We do not have any herbs specific to the Triple Warmer however because the main use of TW.5 is to clear heat/temprature from the whole body, then we could see Huang Qin as a herbal parallel for this point.

There is however a famous ancient formula called *the five flowers* - it is used to clear heat from the whole body;

- Jin Yin Hua
- Pu Gong Ying
- Lian Qiao
- Zi Hua Di Ding
- Ju Hua
- Gan Cao is added to balance the formula

6.11 The Gall Bladder meridian

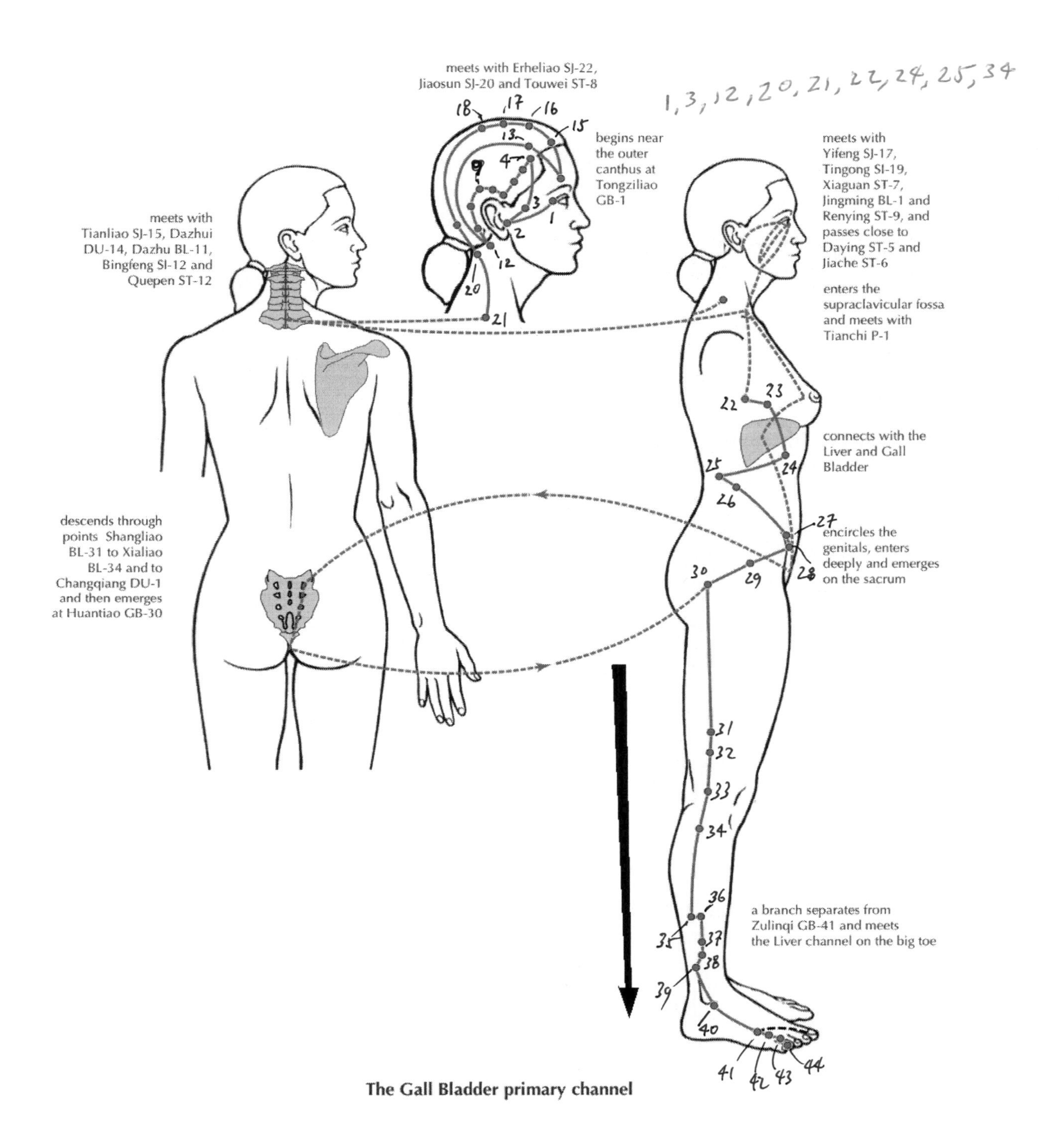

The Gall Bladder primary channel

Illustrations reprinted from A Manual of Acupuncture by Peter Deadman & Mazin Al-Khafaji with Kevin Baker by kind permission of the publishers, www.jcm.co.uk.

Discussion

The Liver and the Gall Bladder are linked physically and via their meridians and some of the Gall Bladder points are used to influence the Liver. The Gall Bladder stores and excretes the bile produced from the old blood of the Liver into the intestines to aid digestion.

6.11.1 The Gall Bladder points used in clinical practice

GB.2

Length of needle used : 1.0"
Depth of insertion : 0.5"
Location : In the depression anterior to the auricular intertragic notch when the mouth is open.
Action : Used reducing for pain, swelling, congestion and inflammation of the ear. Used reinforcing for loss of hearing. This point has a vertical insertion due to the local anatomy. We would get a reducing or reinforcing effect through the acupuncture autoscope machine. Moxa can be applied to the handle of the needle. This point is often combined with TW.21 and TW.3.

How to insert and remove GB.2: we ask the patient to open their mouth wide, we then insert the needle, they then close their mouth. To remove the needle we ask the patient to open their mouth wide, we then remove the needle, they then close their mouth.

GB.14

Length of needle used : 1.5"
Depth of insertion : 1.0"
Location : In the frontal region, 1" superior to the peak of the eyebrow in the depression aligned with the pupil when the eyes are focused forward. We insert the needle at GB.14 and punch it so that the tip of the needle reaches almost to extra point Yu Yao.
Action : This point is punched in one direction only for anatomical reasons. We would get a reducing or reinforcing effect through the acupuncture autoscope machine. Moxa can be applied to the handle of the needle. This point can be used for all eye problems and is used reducing for headache. This point clears wind attack not only from the forehead but also strengthens the ability of the upper back to resist wind attack. We have a Length of needle used = 1.5" and a Depth of insertion = 0.5" because we need to have the needle away from the face when we apply moxa sticks to the end of the needle. This point, when combined with Yin Tang (extra point) reducing helps calm the mind to reduce stress.

GB.20

Length of needle used : 1.0"
Depth of insertion : 0.5"
Location : In the depression between the origins of Sternocleidomastoid and Trapezius muscles. This point is punched towards the opposite eye for an eye problem, or opposite ear for an ear problem due to the local anatomy.
Action : Only used reducing for pain, swelling and inflammation in the local occipital area. Also used reducing to clear the whole head in conjunction with BL.10 for headaches, migraine and wind, cold, damp, attack to head. Moxa can be applied to the handle of the needle.

GB.21

Length of needle used : 1.0"
Depth of insertion : 0.5"
Location : In the suprascapular region, midway between the tip of the acromion process and below the spinous process of the seventh cervical vertebra (C7).
Action : With this acupuncture point we always insert the needle at a 25 degree angle in the direction of the acromion (shoulder) due to the local anatomy, although this is the reinforcing direction we only use this point reducing. We would get a reducing effect through our needle manipulation. This point is used to clear a local tightness and stiffness of the shoulder and neck. It is also a very powerful point for creating a downwards movement in the body. If a woman was in labour this point could be used to bring the baby down and out. So because of this effect the point should not be used on pregnant women because it could cause a premature birth. Also the patient must be lying down when they are having acupuncture on this point or they could faint.

GB.30

Length of needle used : 2.5", or 3" for a larger person.
Depth of insertion : 2", or 2.5" for a larger person.
Location : In the gluteal region, one third the distance from the greater trochanter to the sacral hiatus.
Action : Used reducing for pain, swelling and inflammation of the hip and buttock and sciatica. Moxa can be applied to the handle of the needle.

GB.31

Length of needle used : 1.5"
Depth of insertion : 1.0"
Location : On the midline of the lateral thigh, 7" superior to the popliteal crease. GB.31 can also be located directly inferior to the tip of the middle finger when patient is standing and the arm is extended along the thigh.

Action : Used reducing for pain, swelling and inflammation of the thigh, sciatica. Moxa can be applied to the handle of the needle.

GB.33

Length of needle used : 1.5"
Depth of insertion : 1.0"
Location : On the lateral side of the knee, superior to the jointline, in a depression between biceps femoris tendon and the lateral condyle of the femur.
Action : Used reducing for pain, swelling and inflammation of the knee and sciatica. Moxa can be applied to the handle of the needle.

GB.34

Length of needle used : 1.5"
Depth of insertion : 1.0"
Location : On the lateral side of the leg, in the depression anterior and inferior to the head of the fibula.
Action : Used reducing for pain, swelling and inflammation of the leg and sciatica. This is the influential point of the tendons, so if the patient has a tendon problem (tight tendons, tendonitis, cramping and spasms of the tendons) anywhere in their body, we use this point. This point is also used to reduce any pain, swelling and inflammation or congestion in the MBC relating to the Liver and Gall Bladder. So we reduce for food poisoning, abdominal bloating, inflammation, abdominal spasm and cramping. Moxa can be applied to the handle of the needle.

GB.37

Length of needle used : 1.0"
Depth of insertion : 0.8"
Location : On the lateral side of the leg, 5" superior to the prominence of the lateral malleolus.
Action : Used reducing for pain, swelling and inflammation of the eyes and for any eye problems, the name of this point translates as bright light, so we can clear excessive pathogens, wind/heat etc so the eyes can see clearly and shine brightly again.

GB.38

Length of needle used : 1.0"
Depth of insertion : 0.8"
Location : On the lateral side of the leg, 4" superior to the prominence of the lateral malleolus.

Action : Only used reducing for migraine and headaches, particularly those caused by wind attack to the head.

GB.39

Length of needle used : 1.0"
Depth of insertion : 0.8"
Location : On the lateral side of the leg, 3" superior to the prominence of the lateral malleolus.
Action : Used reducing for pain, swelling and inflammation of the leg, mainly used to treat sciatica. The whole outside of the leg can be treated from this point because this is the meeting point of the three *yang* meridians on the outside of the leg, the Gall Bladder, Stomach and Bladder meridians. As these three meridians also go to the head, this point could support GB.38 in reducing headaches and migraine. When this point is used reinforcing it will strengthen weak legs in old age . This is the also the influential point of marrow so it can help to treat bone problems and blood problems, anemia (the blood is made in the bones). Moxa can be applied to the handle of the needle.

GB.40

Length of needle used : 1.5"
Depth of insertion : 1.0"
Location : At the lateral ankle, in the depression anterior and inferior to the lateral malleolus.
Action : Used reducing for pain, swelling and inflammation of the ankle and can be used reducing to smooth the flow of blood and *qi* through the Liver and whole body. Used reducing to clear blood stagnation caused by traumatic injuries. Moxa can be applied to the handle of the needle for local ankle problems.

GB.41

Length of needle used : 1.0"
Depth of insertion : 0.5"
Location : On the dorsum of the foot between the 4th and 5th metatarsal bones, 1.5" above the web of the fourth and fifth toes. We lift the tendon towards the big toe and insert the needle under the tendon and over the bone.
Action : Only used reducing to clear heat / temperature from the whole body, febrile diseases, temperature, low grade fever and heat causing a rash. This point is often combined with TW.5, they are paired trigram points both used reducing to clear heat from the whole body.

6.11.2 Local Extra Points

Dan Nang Xue

Length of needle used : 1.5"
Depth of insertion : 1.0"
Location : 1" or 2" below GB.34 on the right leg only (because the Gall Bladder is on the right side of the body), the point is at the most tender spot.
Action : This is an extra point used with strong reducing manipulation to break down Gall Bladder stones and reduce Gall Bladder inflammation. Moxa can be applied to the handle of the needle.

6.11.3 Gall Bladder herbs

Herb	Actions
Yan Hu Suo, Yu Jin, Chai Hu	To clear *qi* stagnation in the Gall Bladder.
Huang Qin, Chai Hu, Lian Qiao, Xia Ku Cao, Bai Shao Yao	To clear heat from the Gall Bladder.

bai shao yao, 白芍藥 shao yao 白芍
chai hu 柴胡
huang qin 黃芩
lian qiao 連翹
yan hu suo 延胡索
yu jin 鬱金
xia ku cao 夏枯草

6.12 The Liver Meridian

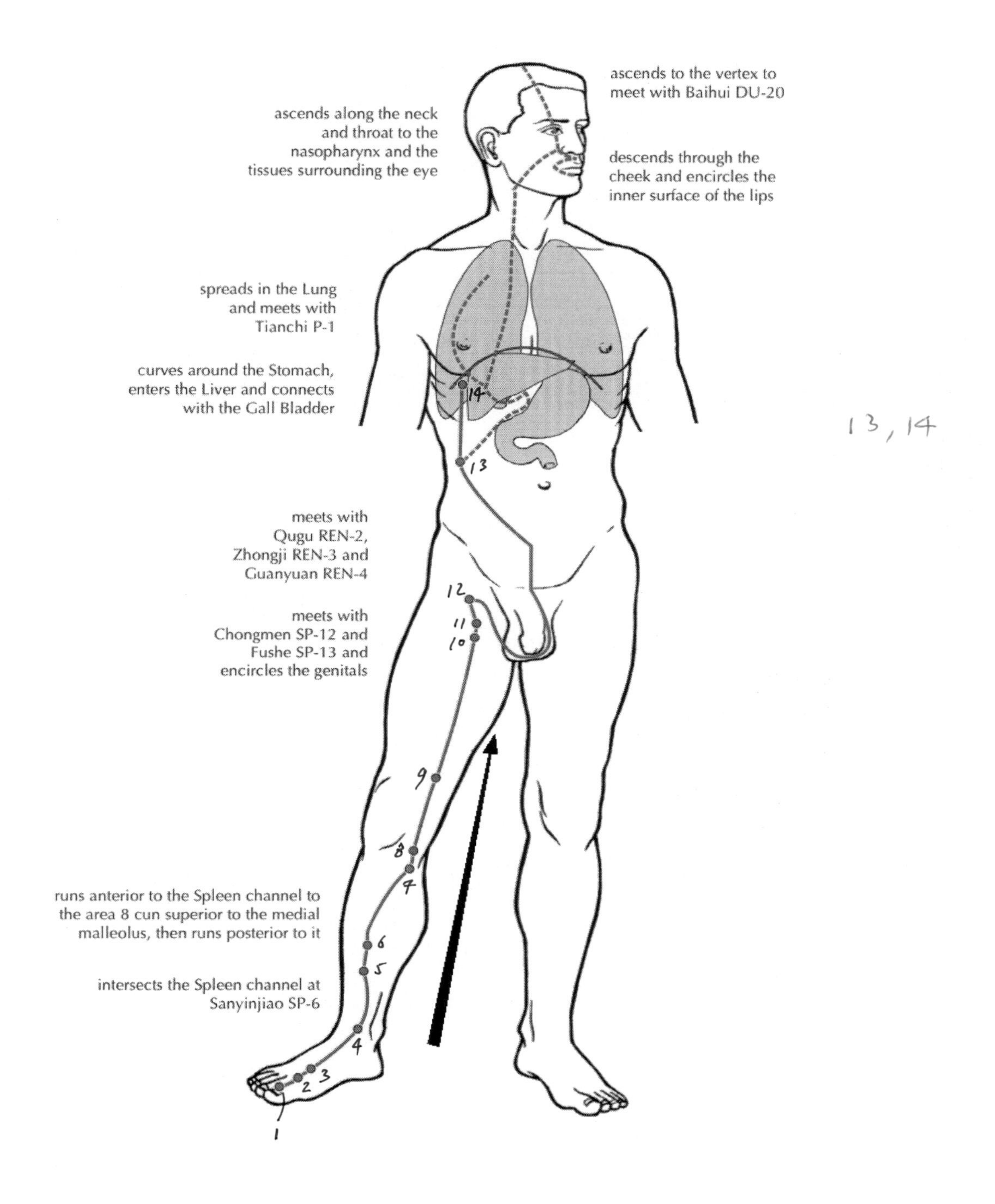

13, 14

Illustrations reprinted from A Manual of Acupuncture by Peter Deadman & Mazin Al-Khafaji with Kevin Baker by kind permission of the publishers, www.jcm.co.uk.

Discussion

The Liver is situated on the right side of the Middle Body Cavity. The Liver meridian connects with the Gall Bladder meridian. The Gall Bladder organ is attached to the Liver. The Liver stores and regulates the blood volume in the body, it is a reservoir of the blood. More blood is released in times of activity and less at times of rest, the blood returns to the Liver at night which is why it is good to sleep on ones right side.

The eyes are the opening of the Liver, if the Liver blood is deficient, the eye sight will be weak if the Liver *yin* is deficient the eyes will be dry. If heat is attacking the Liver there will be redness and pain of the eyes, if there are toxins in the Liver there will be itchiness and inflammation of the eyes/eye lids. If the scellera are yellow there is heat in the Liver.

The Liver controls the tendons, so insufficient Liver blood could lead to tight tendons and weak tendons that could more easily be vulnerable to sports injuries.

If heat is attacking the Liver it could lead to contractions of the tendons/convulsions/spasms

The Liver also controls the nails so if the Liver is not nourished with blood and essence then the nails will be weak. If the Liver blood is strong then the nails will be strong.

The Liver can become unbalanced by being affected by emotions. Also if there is excessivness in the Liver it could lead to emotional problems developing like anger and irritability. There is also a connection between the Liver and the central nervous system.

One of the unseen pathways of the Liver meridian ascends to the top of the head to meet the point GV.20 so wind, heat or *yang* excess from *yin* deficiency in the Liver could cause disturbances to the head resulting in headaches or migraines. The Heart also has an unseen pathways that goes to the head, so Heart and Liver blood deficiency combined with a wind attack to the head can cause vertigo, light headedness and dizziness. The Liver particularly dislikes wind which will affected it more in spring. For this we use Chuan Xiong, Jing Jie, Bai Zhi and Dang Gui.

The Liver filters the blood and clears infections and inflammation so if a person has bruising or swelling or infections and inflammation anywhere in the body we not only treat the local area but also use LIV.3 reducing.

Menstrual problems are not only to do with the genital urinary system and the hormones but also the Liver blood. The Liver meridian also goes through the GU system. When we are treating excessive PMS, we acupuncture LIV.5 reducing.

The hormones flow in the blood so when we are treating hormone related conditions like premenstrual tension with breast distention we treat the Liver as well.

The Gall Bladder secretes bile which helps in the process of digestion and absorption of food is attached to the Liver. So the Liver is therefore involved with the Spleen and Stomach and their processes of digestion and absorption of food. Because of this connection if there is an imbalance in the Middle Body Cavity the Liver is often also treated. Stagnated Liver energy can result in abdominal fullness and irritability.

6.12.1 The Liver Points used in Clinical Practice

LIV.1

Length of needle used : 0.5"
Depth of insertion : 0.2"
Location : On the lateral side of the first digit, the big toe, 0.1" from the corner of the nail bed.
Action : Used reducing for arthritis or pain, inflammation or swelling along the big toe. For this moxa can be applied to the handle of the needle. To reduce hyper-menorrhoea (excessive menstrual bleeding or mid cycle bleeding) we SB and then apply 3 Δ to this point in conjunction with SB and apply 3 Δ on SP.1. This combination would be used after menstruation had finished.

LIV.3

Length of needle used : 1.0"
Depth of insertion : 0.5"
Location : On the dorsum of the foot, between the first and second metatarsal bones, approximately 1.5" superior to the web margin.
Action : Used reducing to clear Liver heat, pain, swelling and inflammation, also to clear inflammation anywhere in the body and traumatic injuries. This point will also help to smooth the blood and *qi* flow so it is used to regulate menstruation. This point is also used for eye problems because the eyes are the opening of the Liver, for this it would be combined with GB.37. The unseen pathway of the Liver goes to the top of the head so this point can be used to reduce headaches and irritability. We can also use this point with a vertical insertion to nourish the Liver blood and Liver *yin*. This point is not used reinforcing. No moxa used on this point.

LIV.5

Length of needle used : 1.5"
Depth of insertion : 1.0"
Location : On the medial side of the leg, posterior to the tibial medial margin, 5" superior to the medial malleolus.
Action : This point can be used reducing prior to menstruation to reduce abdominal pain and cramping, it would be supporting SP.6 reducing. If we reduce this point it will reduce pain, swelling, itching and inflammation in the GU system. When acupunctured in the reducing direction prior to menstruation it can bring down endometriosis (upwards flow of menstrual blood which is attaching to internal organs and other structures). This point can be used reinforcing to strengthen a weak Liver that is recovering from a chronic Liver problem like hepatitis. If we reinforce this point combined with LIV.8, BL.18, SP.6 and KID.3 it will replenish the Liver blood.

6.12.2 Liver herbs

Herb	Actions
Chai Hu	To conduct to the Liver
Bai Shao Yao, Yan Hu Suo, Yu Jin	To relax the Liver and counter contrations and spasms of the tendons.
Xiang Fu, Chai Hu, Qing Pi	To disperse Liver *qi* stagnation.
He Shou Wu, Nu Zhen Zi, Dang Gui, Chuan Xiong, Ji Xue Teng	To replenish Liver blood deficiency.
Nu Zhen Zi, Gou Qi Zi	To replenish Liver *yin*/essence deficiency.
Mu Dan Pi, Chi Shao Yao, Tao Ren, Dan Shen (Miltiorrhizae)	To clear Liver blood stagnation.
Xia Ku Cao, Chi Shao Yao, Huang Qin,	To clear Liver heat and inflammation.
Huang Qi, Tu Si Zi, Nu Zhen Zi,	To strengthen a weak Liver.
Sang Ji Sheng, Niu Xi, Gou Ji	To strengthen the Liver to strengthen the tendons.
Ju Hua, Xia Ku Cao, Sang Ye, Mi Meng Hua	To clear eye inflammation.

chai hu 柴胡
chi shao yao 赤芍藥
chuan xiong 川芎
dan shen 丹參
dang gui 當歸
gou ji 狗脊
gou qi zi 枸杞子
he shou wu 何首烏
huang qi 黃耆 黃芪
huang qin 黃芩
ju hua 菊花
mi meng hua 密蒙花
mu dan pi 牡丹皮
niu xi 牛膝 懷牛膝
nu zhen zi 女貞子
qing pi 青皮
sang ji sheng 桑寄生
sang ye 桑葉
shao yao 白芍 bai shao yao, 白芍藥
tao ren 桃仁
tu si zi 菟絲子
xia ku cao 夏枯草
xiang fu 香附
xue teng 血藤
yan hu suo 延胡索
yu jin 鬱金

6.13 The Conception meridian/Vessel

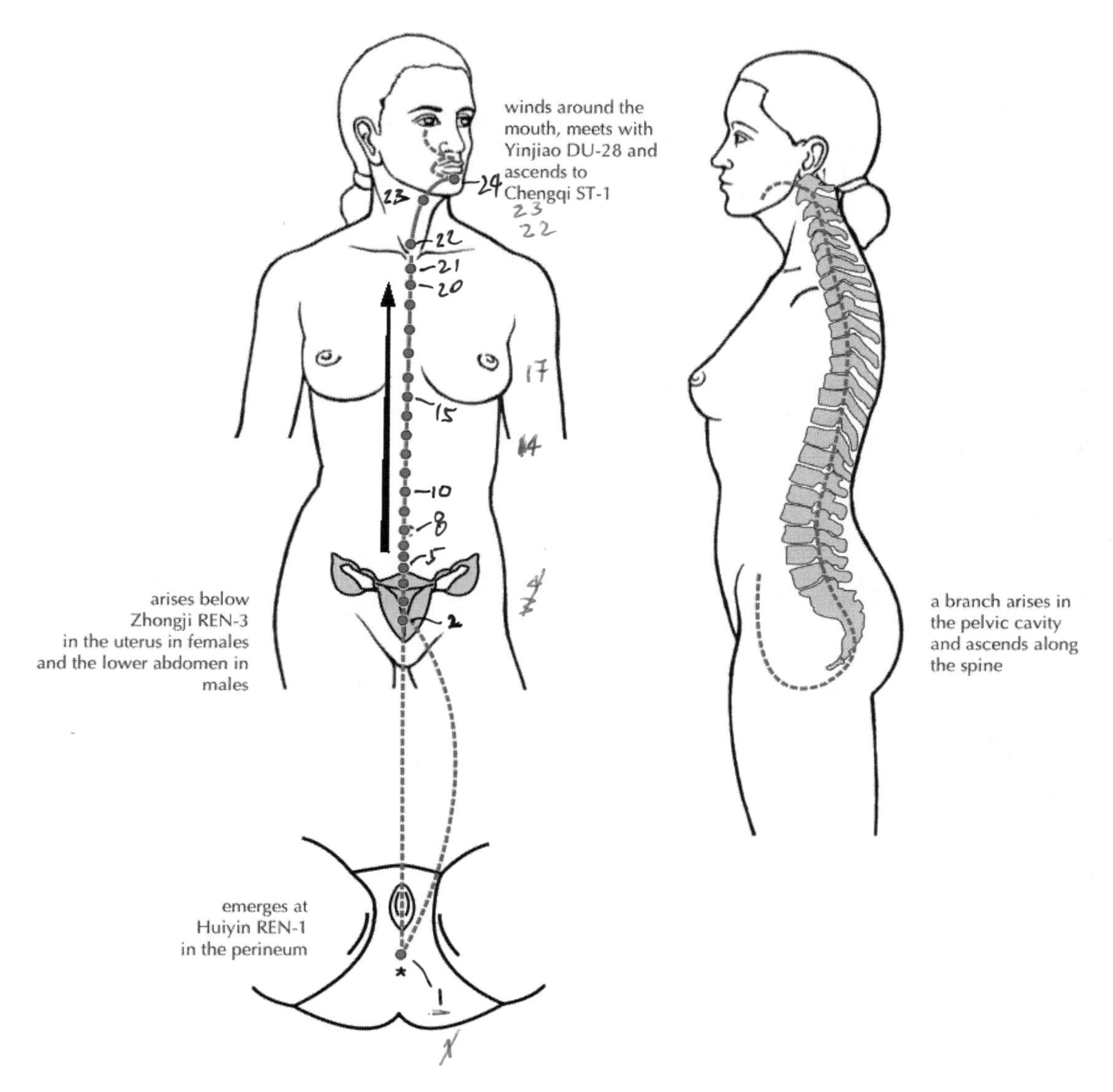

The Conception vessel primary pathway

Illustrations reprinted from A Manual of Acupuncture by Peter Deadman & Mazin Al-Khafaji with Kevin Baker by kind permission of the publishers, www.jcm.co.uk.

Discussion

The conception meridian regulates and supplies all the *yin* meridians in the body, it runs along the midline of the front of the body, from the perineum all the way up the torso and neck to end on the middle of the lower gum.

If the *qi* in this meridian is flowing freely then it is able to supply all the *yin* meridians in the whole body.

The Conception meridian is thought of as a huge lake of water that can flow down all the *yin* meridians and nourish all the internal organs. It is the *yin* meridians that are more concerned with nourishing and regenerating the body. It is sometimes called the conception vessel because it holds the *yin* essence like a vessel would hold water.

There are many Front Mu points on the Conception meridian that are directly over various internal organs and so can be used to control their activity.

6.13.1 The conception vessel points used in clinical practice

CV.3 – *Bladder Mu*

Length of needle used : 1.0"
Depth of insertion : 0.5"
Location : In the pubic region, on the midline, 4" below the navel.
Action : Only used reinforcing to strengthen LBC, fertility, improve sexual potency, replenish essence and strengthen Kidneys, Bladder and GU system. This point is often combined with CV.4 and KID.12.

CV.4

Length of needle used : 1.0"
Depth of insertion : 0.5"
Location : In the pubic region, on the midline, 3" below the navel.
Action : Only used reinforcing to strengthen LBC, fertility, improve sexual potency, replenish essence and strengthen Kidneys, Bladder and GU system. CV.4 is the meeting point of all *yin* meridians in the body so it is a good point to replenish the bodies essence. This point is often combined with CV.3 and KID.12.

CV 6 – *Sea of qi*

Length of needle used : 1.0"
Depth of insertion : 0.5"
Location : In the pubic region, on the midline, 1.5" below the navel.
Action : Only used reinforcing to strengthen the *qi* and *yang*/functional power of the whole body.

CV 12 – *Stomach Mu*

Length of needle used : 1.5"
Depth of insertion : 1.0"
Location : On the midline, 4" above the navel.
Action : Used reducing for pain, swelling, congestion, distention and inflammation of the MBC and Stomach, nausea, vomiting. This point is often combined with CV.13, ST.21, KID.19. Moxa can be applied to the handle of the needle.

CV.13

Length of needle used : 1.0"
Depth of insertion : 0.5"
Location : On the midline, 5" above the navel.
Action : Used reducing for pain, swelling, congestion, distention and inflammation of the MBC and Stomach, nausea, vomiting. This is a support point for CV.12. This point is often combined with CV.12, St.21, KID.19. Moxa can be applied to the handle of the needle.

CV.17

Length of needle used : 1.0"
Depth of insertion : 0.5"
Location : On the high point of the sternum on the midline, between the nipples.
Action : Used reducing for pain, swelling, congestion, distention and inflammation of the Lungs and chest problems. It relaxes and opens up and calms the chest. This point is needled reducing when combined with LU.1 and KID.27 reinforcing. CV.17 has a very shallow degree of insertion and is almost flat against the skin. This is for two reasons, first the sternum (breastbone) is just under this point and also because there is a rare condition called sternal foramen (a hole in the sternum) and to avoid going through this in to the thoracic cavity and causing pneumothorax we have a very flat angle of insertion for the acupuncture needle.

CV.22

Length of needle used : 1.0"
Depth of insertion : 0.5"
Location : In the suprasternal fossa, on the midline, just above the jugular notch/pit of the throat.
Action : Used reducing for pain, swelling and inflammation of the throat and to clear phlegm stuck in the throat or causing a congested and blocked feeling in the throat, it relaxes and opens the throat. Lift skin and insert at a shallow angle. Moxa can be applied to the handle of the needle.

6.14 The Governing Meridian/Vessel

14, 15, 16, 26

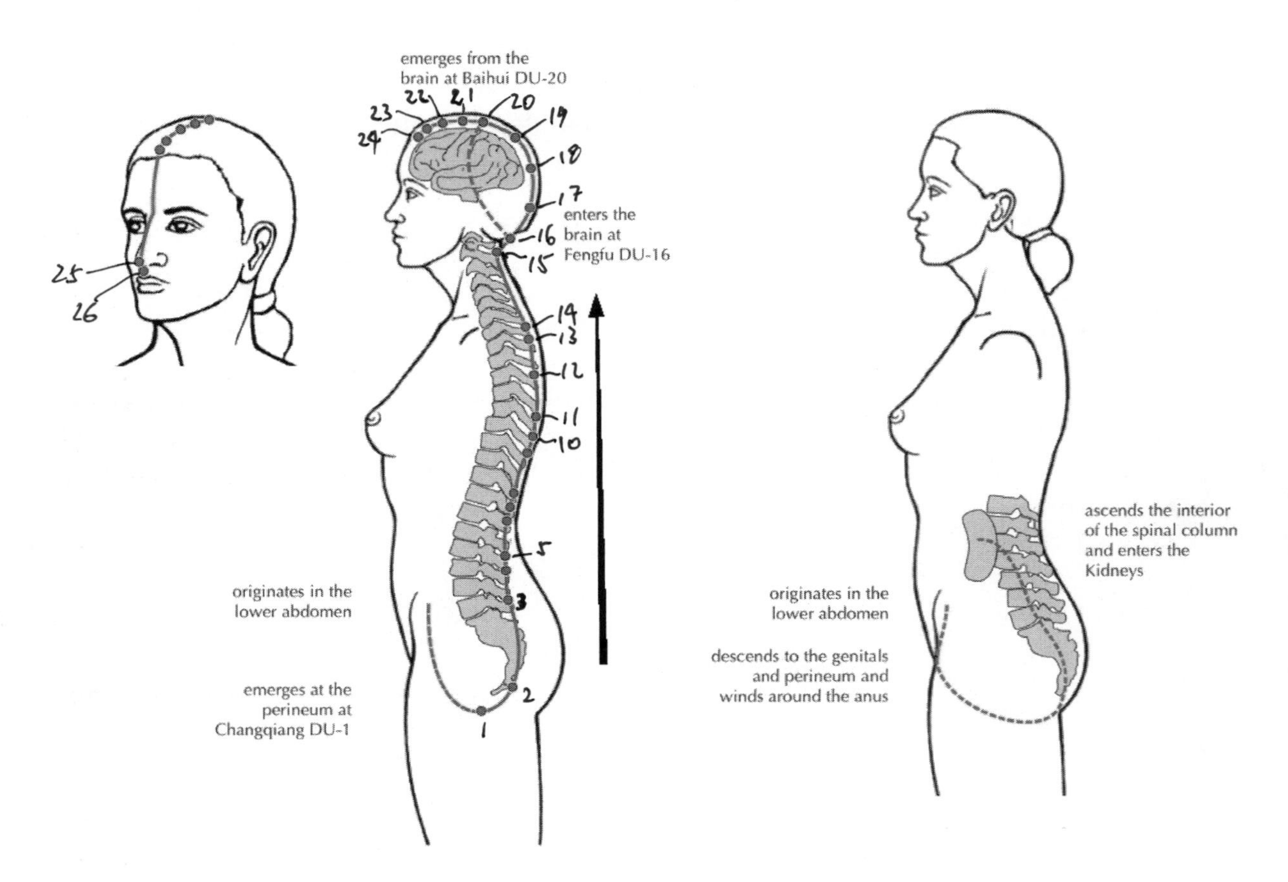

The Governing vessel primary pathway

The Governing vessel first branch

Illustrations reprinted from A Manual of Acupuncture by Peter Deadman & Mazin Al-Khafaji with Kevin Baker by kind permission of the publishers, www.jcm.co.uk.

Discussion

The Governing meridian regulates and supplies all the *yang* meridians in the body. Its pathway runs along the midline of the back of the body, from the coccyx all the way up the spine, over the head and ends on the middle of the upper gum. If the *qi* in this meridian is flowing freely then it is able to supply all the *yang* meridians in the whole body with *qi*.

The Governing meridian is thought of as a sun shining down radiating sunlight along all the *yang* meridians providing functional power to all the internal organs.

The *yang* meridians are more concerned with defending the body against attack from Exogenous Pathogenic Factors like viruses, bacteria and extreme weather conditions. If heat has already attacked the body/spine we can clear it by reducing GV.14 (just below the 7th cervical vertebra) because it is the meeting point of all *yang* meridians in the UBC.

The Governing vessel is strengthened directly by the Kidneys which control the bones, so if the Kidneys are weak the governor vessel will be weak. If we want to strengthen the spine we reinforce BL.23 and apply moxa. This will improve the brain function, the 'willpower' and determination will be strengthened and the person will be more positive, more focussed, more courageous and decisive.

All the nerves in the whole body have their roots in the spine and the nerves control the whole body so a healthy spine is very important for our general health.

Aching or soreness below lumbar vertebrae 4 and 5 indicate a very depleted spine/GV/Kidneys this weakness means the spine can easily be injured or put out and if injured, it can often take a long time to heal. So when soreness is felt in the lumbar area of the spine, reinforce BL.23 and moxa.

6.14.1 The Governing vessel points used in clinical practice

The points on the GV are listed in reverse order, to make it easier to explain their uses.

GV.26

Length of needle used : 0.5"
Depth of insertion : 0.3"
Location : One third of the distance from the nose to the top of the upper lip.
Action : If there is acute pain in the spine, then before we acupuncture the spine we quick punch GV.26 with a half inch needle to send a blast of energy straight down the central nervous system (CNS) through the centre of the spine to clear the channel. To acupuncture GV.26 we puncture through the upper lip towards the upper gum. GV.26 is also an emergency revival point for unconsciousness.

GV.23

Length of needle used : 1.0"
Depth of insertion : 0.5"
Location : On the midline, 1" posterior to the anterior hairline.

Action : Used reducing for pain, swelling and inflammation of the nose, eyes and for frontal headache.

GV.20

Length of needle used : 1.0"
Depth of insertion : 0.5"
Location : On the midline, at the intersection of a line connecting the apex of the right and left ears.
Action : Used reducing for headache on the top of the head. Used reinforcing then remove the needle and apply 3 Δ to lift up prolapse/collapse of the internal organs, combined with SP.6 reinforcing and LI.4 reducing. For dizziness caused by wind attack to the head with blood deficiency in the head we insert the needle in the reducing direction and apply moxa to the handle of the needle.

GV.16

Length of needle used : 0.5"
Depth of insertion : 0.3"
Location : On the posterior head, 0.5" directly below the external occipital protuberance.
Action : Used reducing for pain, swelling and inflammation of the back of the head and neck, clears wind EPA, calms the mind.

GV.14

Length of needle used : 1.5"
Depth of insertion : 1.0"
Location : Below the spinous process of the seventh cervical vertebra (C7).
Action : Used reducing if EPA heat has attacked the UBC body/spine we can clear heat from a wide area with this point because it is the meeting point of all *yang* meridians in the UBC. Used reducing for pain, swelling and inflammation of the neck and shoulders. Stiffness of the spine and neck with inability to turn the head.

The Governing Vessel used in conjunction with the Bladder Meridian Shu Points

To balance the reinforcing of the Back Shu points we reduce the GV point that is at the same level (apart from Spleen Shu).

GV.12

Length of needle used : 1.5"
Depth of insertion : 1.0"
Location : Below the spinous process of T3.
Action : When we reinforce BL.13 (Lung Shu) we reduce GV.12 below T3.

GV.11

Length of needle used : 1.5"
Depth of insertion : 1.0"
Location : Below the spinous process of T5.
Action : When we reinforce BL.15 Heart Shu we reduce GV.11 below T5

GV.10

Length of needle used : 1.5"
Depth of insertion : 1.0"
Location : Below the spinous process of T6.
Action : When we reinforce BL.16 Governing Shu (used to support Heart Shu) we reduce GV.10 below T6

GV.9

Length of needle used : 1.5"
Depth of insertion : 1.0"
Location : Below the spinous process of T7.
Action : When we reinforce BL.17 Diaphragm Shu we reduce GV.9 below T7

GV.8

Length of needle used : 1.5"
Depth of insertion : 1.0"
Location : Below the spinous process of T9.
Action : When we reinforce BL.18 Liver Shu we reduce GV.8 below T9

GV.4

Length of needle used : 1.5"
Depth of insertion : 1.0"
Location : Below the spinous process of L2.
Action : When we reinforce BL.23 Kidney Shu we reduce GV.4 below L2

6.14.2 Local Extra Points

Yin Tang

Length of needle used : 0.5"
Depth of insertion : 0.3"
Location : Midway between the medial ends of the eyebrows.
Action : The name of this point is doorway to the mind, so it can be used reducing to calm the mind, to counter over-thinking. Yin Tang is also often used as a support point for LI.20 for nose problems, see LI.20.

Below T4

Length of needle used : 1.5"
Depth of insertion : 1.0"
Location : Below T4.
Action : When we reinforce BL.14 - Pericardium Shu (used to support Lung Shu), we reduce below T4. There is no GV point here.

Chapter 7

Acupuncture of the Back

7.1 Before acupuncturing the back

If there is acute pain in the spine, then before we acupuncture the spine we quick punch GV.26 with a half inch needle to send a blast of energy straight down the central nervous system (CNS) through the centre of the spine to clear the channel. To acupuncture GV.26 puncture through the upper lip towards the upper gum.

If there is chronic pain in the back, then before we acupuncture the spine we quick puncture BL.40 to clear inflammation from the Bladder channel which runs along the back. To acupuncture BL.40 we do a vertical punch with a one and a half inch needle to a depth of about one inch.

7.1.1 The meridians of the Back

The Governing Meridian

The centre line of the back is the GV and this runs in an upwards direction. 99 % of the time this meridian is acupunctured in the reducing direction, which is downwards towards the coccyx.

The Hua Tuo points

The Hua Tuo (HT) points are half an inch beside the GV and run down, 99 % of the time this meridian is acupunctured in the reducing direction, which is upwards towards the head. Hua Tuo points on the back can be used for neurological problems, traumatic injuries, control of the muscles, skin and the internal organs.

Hua Tuo was famous for his abilities in acupuncture, moxibustion and herbal medicine, his pioneering work with spinal acupuncture led to the points on the back, half an inch beside the midline, to be named after him. Hua Tuo died in about 208 CE.

The Bladder meridian

The first branch of the Bladder meridian is one and a half inches beside the GV and the second line is three inches beside the GV. The direction of flow of the *qi* in the

Figure 7.1: Hua Tuo

Bladder meridian is downwards, 99 % of the time this meridian is acupunctured in the reinforcing direction, which is downwards towards the coccyx.

(The Bladder points on the sacrum are not the same as the Hua Tuo points on the sacrum)

The Bladder meridian points which are 1.5" lateral to the midline are used to strengthen the internal organs because the points are directly on top of the organs. These points are called Shu Points.

The most used back Shu points

- BL 13, below T3 Lung Shu is used for chronic and acute lung problems
- BL 15, below T5 Heart Shu is used for Heart disease
- BL 17, below T7 Diaphragm Shu is used to replenish the blood
- BL 20, below T11 Spleen is used to strengthen the Spleen and immunity
- BL 23, below L2 Kidney Shu is used to strengthen the Kidneys and immunity

7.1.2 How to use the Hua Tuo points

We use Governing Meridian Points with Hua Tuo Points to control the root of the nerve, to control the nerve pathways, to control the blood flow, muscles and the internal organs.

If there is a pain or inflamation causing hyper-activity it could be radiating from an internal organ or it could be muscular or skeletal or neurological. For this very common situation we use the reducing acupuncture method. GV and Hua Tuo Points will be tender if the condition is excessive.

If there is a weakness causing hypo-activity then we use the reinforcing method, this is very rarely used, it would be in cases of post-stroke weakness or to regenerate muscle wasting.

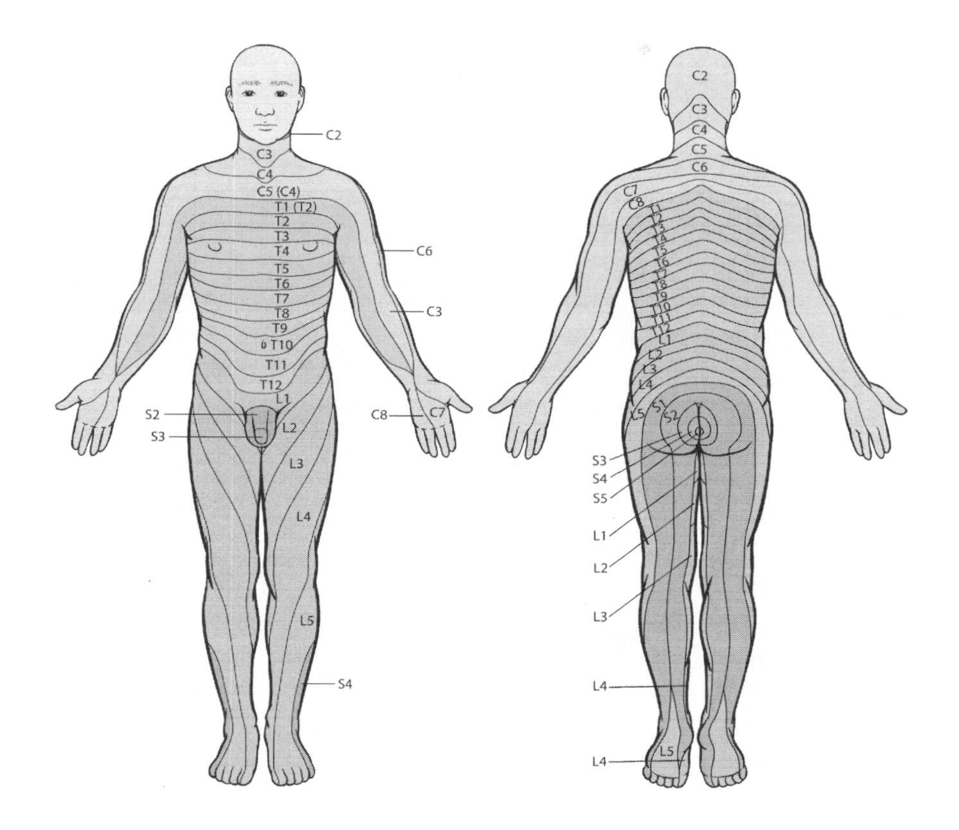

7.1.3 Detailed examples of how to use Hua Tuo Points

Neck problems

For neck problems we use GV and the relevant HT points reducing, it could be from C7 up to C1, consultation and gentle finger pressure are used to diagnose which cervical vertebrae (neck) need to be treated. Local support points that are used are BL.10, GB.20 and GB.21. Also in the formula would be SI.3 and BL.62 both reducing because they are the distant trigram paired points to reduce pain, swelling inflammation and tension in the upper back and neck.

Arm problems

For referred pain in the arm that is really coming from a problem in the neck we use GV and Hua Tuo points in the reducing direction, below C4, C5, C6, and C7 (we only use the HT points on the affected side) to relax the nerves or reduce bone swelling that press on the nerves which can be the cause of numbness or some inflammation that causes pain. We could support this work with relevant acupuncture points along the arm and hand.

Upper back problems

For referred pain in the upper back that is really coming from a problem in the spine we use GV and Hua Tuo points in the reducing direction, below C5, C6, C7, and T1 to relax the nerves or reduce bone swelling that press on the nerves which can be the cause of numbness or some inflammation that causes pain. We could support this work with relevant acupuncture points around the edge of the scapula where you can find muscle knots in the trapezius and rhomboid muscles that you needle against the direction of the Bladder and Small Intestine meridian with the arrowhead formation using two or three 1" needles. Local points that are also often used are SI.9, SI.10, SI.11, SI.12, SI.13, SI.14 and SI.15 also GB.21, and if the shoulders are affected then we use TW.14 supported by TW.13 and LI.15 supported by LI.14 the distant points for the shoulder problem would be LI.4 and TW.5.

Also in the formula would be SI.3 and BL.62 both reducing because they are the distant trigram paired points to reduce pain, swelling inflammation and tension in the upper back and neck. Just the side that is affected would be used, if it is across the whole of the upper back both sides would be used.

Leg problems

For referred pain in the legs that is really coming from a problem in the lower back, sciatica/lumbago, we use use GV points below L 1, L 2, L 3, L 4 and L 5 and S1 and S2 reducing and Hua Tuo points in the reducing direction below L 1, L 2, L 3, L 4 and L 5 and S1 and S2 on the effected side to control the nerves that run through the lower back, pelvis, quads, buttocks, calf muscles and hamstrings. We could support this work with relevant acupuncture points along the leg and foot.

We may need to add the above the hip and the below the hip points (one and a half inch needle, vertical punch)

Pain on the side of the leg

If pain is on the side of the hip we use GB.29 (press and if it twinges this is the point), if the pain is in the buttock we use GB.30 (3" needle slightly against the GB meridian, through the big hole of the pelvis), if the pain is on the side of the leg we use GB.31 (the tip of the patients middle finger lands on this point on the side of leg), we use GB.34 to relax the tendons and we use GB.39 because it is the meeting point of all *yang* meridians on the leg, we also use BL.59 and BL.60.

Pain on the back of the leg

When pain goes down the back of the leg use BL.36 (in the centre of the crease of the buttocks) and BL.37 which is six inches below BL.36.

Kidney points to support treating pain on the back and leg

When we do acupuncture on the lumbar vertebra either for pain in the lumbar vertebra or for pain that runs down the side or back of the leg we always also reinforce BL.23, BL.24 and B.52 and balance this with GV.4 reducing. This is because the Kidneys strengthen the bones in general and the lumbar vertebrae in particular. So this would reduce any bone swelling that was pressing on the nerves and also keep the back strong whilst we did all reducing acupuncture. We also would reinforce KID.3.

For problems with the anus and sexual organs and GU system

The four sacral nerves pass through the sexual organs and the anus, so for any problems with the anus and sexual organs and GU system (e.g. collapse of the anal sphincter, hemorrhoids, piles or if there is *qi* stagnation, inflammation, pain or swelling of the prostrate or sexual organs) we use Hua Tuo points below S 1, S 2, S 3 and S 4 and BL.31, BL.32, BL.33 and BL.34 and the GV points below S 1, S 2, S 3 and S 4 all reducing to clear inflammation and blood stagnation and then we apply moxa to the needles to tonify and lift the collapse and invigorate energy circulation. The sacrum is a big bone and bones are instrumental in blood production therefore we can also use BL.31, BL.32, BL.33 and BL.34 for blood related problems in women (menstrual or hormonal conditions). In men we use these points for poor function of the GU system or for impotence, we reinforce and apply moxa to strengthen the sexual function to counter impotence and enhance fertility.

For problems with the bones and health of the whole body

When you reinforce BL.23, BL.24 and BL.52 (balance this with GV.4 reducing) to strengthen the Kidneys to strengthen the bones in the whole body (arthritis, osteoprosis etc) or for poor general health, weak kidney function, insufficient kidney essence, poor sexual function, or weak Kidneys causing the lumbar vertebrae to be weak and aching or have a hollow feeling or a cold feeling in the Kidneys. You can support this work by adding BL.31, BL.32, BL.33 and BL.34 reinforcing. We also add BL.11 reinforcing, the feature point for bones.

For problems with the internal organs

When we use acupuncture on the back to treat the internal organs we most often strengthen the the lung, Heart, Spleen and kidney by reinforcing the back Shu points on the Bladder meridian and to treat the Liver, Gall Bladder and Stomach we reduce Hua Tuo's points below T 9, T 10 and T 12.

7.2 Common acupuncture formulas on the back for general health problems

Formula to regulate the Heart and Kidneys

This is a useful formula to reduce stress and balance spiritual and emotional problems.

The Heart is fire and the Kidneys are water. Life can cause people to have depletion of kidney water/essence and excess Heart fire. Kidney water deficiency results in weak legs, knees, sexual function, lack of ambition etc. Heart fire excess results in restlessness, over thinking, Heart pain, high blood pressure, overheating etc. Also the less water the Kidneys have the less its ability they have to cool down the excess Heart fire. The more fire the less water etc.

So to replenish kidney essence deficiency we reinforce BL.23, BL.24 and BL.52 and balance this with GV.4 reducing. To clear excess Heart fire we reduce GV below T5 and T6 and Hua Tuos points below T5 and T6.

The distant support points would be KID.3 reinforcing and this would be balanced with BL.60 reducing. Also LIV.3 reducing to harmonise the soul in the Liver and reduce irritability and also LI.4 reducing to calm the thinking mind in the brain and H.7 vertical punch to harmonise the emotional mind in the Heart. Also we could reduce the extra point located midway between the medial ends of the eyebrows, this point is translated as gateway/door to the mind/spirit, this would calm the mind/spirit. We could support this with GB.14.

If the patient has weak immunity then we could add the formula to strengthen the immune system/Spleen.

Formula to strengthen immune system

If the patient has weak immunity then we could add in BL.20 reinforcing which is Spleen Shu below T11. (We do not use the GV point below T11 when we acupuncture BL.20). The distant support point to strengthen the Spleen would be SP.6 reinforcing. If the patient has irritability then we could add the formula to reduce the LIV and GB.

Formula to reduce the Liver and Gall Bladder

The Liver connects to the eyes and its unseen pathway goes right through the centre of the brain up to the top of the head, if there is excess heat in the Liver or Liver wind rising this can cause irritability so we reduce the Liver by reducing GV and HT below T9. The Gall Bladder meridian zig zags over both sides of the head and temples, so headaches, stress and migraines are often due to Gall Badder excess so we reduce GV and HT below T10.

The distant support point to smooth GB *qi* flow is GB.40 reducing with a one and a half inch needle. The distant support point to relax the tendons in the whole body to release tension and tightness is GB.34 reducing with a one and a half inch needle.

Formula to clear wind and EPA from the Lungs

First we acupuncture BL.11 and BL.12 reducing, then we remove the needles and cup these points. If there was heat in the UBC then we could reduce GV.14 below C7. GV.14 is the meeting point of all the *yang* meridians in the UBC.

The distant support points to reduce lung EPA and support the Kidneys to support the Lungs is LU.7 reducing and KID.6 reinforcing, this is one of the eight trigram paired points.

Drag cupping on the back to clear heat in the UBC

If a patient has massive heat trapped in the UBC, (their skin will come up with a red welt when the upper back is pressed) then we can use drag cupping to clear off the heat.

We cover the back with a very thin layer of oil so the cup can move, then we create a vacuum in the cup and place it onto GV.14 and then drag it down the Governing vessel as far as T6, we then go up the right one and a half inch branch of the Bladder meridian, then across to GV.14, down the Governing vessel to T6 and then up the left one and a half inch branch of the Bladder meridian. We then repeat the same procedure but this time with the three inch branch of the Bladder meridian.

We then make outwards facing crescent moon shapes by dragging the cup along the medial sides of the scapula, this area is where knots in the trapezius and rhomboid muscles are found.

We finish by returning the cup to GV.14 and then we remove it.

The distant support points to reduce redness on the skin is LU.10 because the skin is the third lung, we breath with the Lungs and through the skin. We also reduce LIV.3 to clear the heat from the blood because the Liver stores the blood. Also SB from LI.1, LU.11, LIV.1, SP.1 and SP.10 and the extra point one inch above SP.10.

Chapter 8

Small Acupuncture formulas used in Clinical Practice

Although many of these have been previously mentioned, these small formulas are all gathered here together for further study.

8.1 The Eight Trigram Points

- **L.7 Reducing, KID.6 Reinforcing :** Used for lung conditions i.e. exogenous pathogenic attack, clear heat or wind from the lung, asthma, bronchitis. Additionally used for clearing heat caused by late nights
- **P.6 Reducing, SP.4 Reinforcing :** Used for Heart problems, digestive problems and diabetes
- **TW.5 Reducing, GB.41 Reducing :** Used to clear heat from the whole body caused by exogenous pathogenic attack, late nights, stress or alcohol poisoning
- **SI.3 Reducing, BL.62 Reducing :** Used to clear pain and inflammation from upper back, neck and shoulders

8.2 The Additional Paired Points

These are a series of small formulas using LI.4 which are used extensively in clinical practice.

- **LI.4 Reinforcing, SP.6 Reducing :** This combination will create a downward movement of *qi* in the body. Used clinically for amenorrhoea (no period), and for dysmenorrhea (painful periods). LIV.5 reducing can be added to this formula.
- **LI.4 Reducing, SP.6 Reinforcing :** This combination will create an upwards movement of *qi* in the body. Used for deficiency patients who are weak with low immunity, used for blood deficiency in head causing light headednes, fainting, dizzy, vertigo. Used in case of organ (visceral) prolapse i.e. Stomach, intestines, anus, uterus. Additionally, GV.20 with direct moxa and at least 3 moxa cones.

- **LI.4 Reducing, KID.7 Reinforcing :** Used to stop excessive sweating due to a deficiency condition, menopausal symptoms, spontaneous sweating and night sweating.
- **LI.4 Reducing, ST.36 Reducing :** Used for indigestion, abdominal distention, bloating, swelling, food stagnation, food poisoning. If accompanied by severe diarrhoea, add ST.37. If abdominal swelling and pain is severe, add the following points reducing; GB.34, LIV.3, CV.12 and SP.6.

8.3 Spleen 6 Combinations

It is important to understand the many uses of this point. It is used for a wide variety of conditions and listed here are the most used combinations. SP.6 has such a strong strengthening effect on the whole body because the kidney and Liver meridians also pass through it. Its known as the 'three *yin* intersection'. We can use this point in combination with KID.3 and BL.23 (Kidney Shu) to rehydrate and replenish the *yin*, or essence, of the body.

- **SP.6 Reducing, LI.4 Reinforcing :** This combination will create a downward movement of *qi* in the body so it can be used prior to menstruation for pre-menstrual abdominal cramping and pain, endometriosis, amenorrhoea and to induce labour.
- **SP.6 Reinforcing, LI.4 Reducing :** This combination will create an upwards movement of *qi* in the body. Used for deficiency patients who are weak with low immunity. Used for blood deficiency in head causing light headednes, fainting, dizziness and vertigo. Used in case of organ (visceral) prolapse i.e. Stomach, intestines, anus, uterus. Additionally, GV.20 with direct moxa (several cones).
- **SP.6 Reducing, LI.4, ST.25, ST.26, ST.28, ST.36, ST.37, ST.44, GB.34 & LIV.3 Reducing :** This combination is used for pain, inflammation, swelling of the LBC relating to the digestive system. The abdominal swelling could be caused by wind leading to abdominal bloating, food stagnation or food poisoning causing inflammation. Additionally an EPA to the LBC such as gastritis or gastroenteritis.
- **SP.6 Reinforcing, KID.3, KID.6, ST.36, KID.12, CV.3, LU.1 & KID.27 Reinforcing :** To strengthen the Spleen, Liver, Stomach, Lungs and Kidneys. This will strengthen the blood, essence and *qi* of the whole body. To avoid causing an excess condition from this prescription, we balance these points by reducing LI.4, BL.60, GB.40, GB.34, LU.7, LIV.3 & CV.17.

Chapter 9

Needle technique

Students should practice needle technique for fifteen minutes every day on a cushion. There are 7 methods to practice.

1. Punching: use focus, concentrated effort, intention, speed and decisiveness.
2. Lifting and Thrusting Reducing Method: one, two, three, in and one out
3. Lifting and Thrusting Reinforcing Method: one in and one, two, three out
4. Rotating and Twisting Reducing Method
5. Rotating and Twisting Reinforcing Method
6. Reinforcing Method Lifting & Thrusting combined with Rotating and Twisting
7. Reducing Method Lifting and Thrusting combined with Rotating and Twisting

Method 7 is used to reduce excessive conditions and also to relax the space between the vertebrae when we are inserting needles in to GV and HT points. Methods 1 and 7 are the most important because they are used the most in clinical practice.

9.1 Details about Punching the needle

If we grow the index fingernail and mark the skin with an X this will break the surface tension of the skin and make it easier to insert the needle. Then place your little finger on the skin just to the side of the X, this is to distract the bodies resistance in the skin, as the patients attention is now sent to the area where the little finger is applying pressure and not the area to be punctured and marked by the X.

The initial puncture goes through the skin as quickly and smoothly as possible so as not to unnecessarily irritate the nerve endings on the skin surface. Once the needle has been punched through the skin surface to a depth of about 2 or 3 mm then it is threaded to the correct depth of half or one Chinese inch deep (or cun).

Punching and Threading are different, to get through the skin surface use Punching not Threading. This is because Punching is quick and Threading is slow. Puncture has to go through the skin as quickly as possible so as not to unnecessarily irritate the nerve endings on the skin surface. When we use Threading it is done more slowly with great sensitivity, because we are feeling our way and so can find the right pathway to insert the needle.

If the needle is pointing against the direction of the *qi* flow in the meridian it will have a reducing effect. If the needle is pointing in the direction of the *qi* flow in the meridian it will have a reinforcing effect.

9.2 Details about inserting and withdrawing the needle

For the reducing needle technique: Insert in 3 steps against the direction of the meridian and to take it out, stretch the skin and withdraw in 1 step. For the reinforcing needle technique: Insert in 1 step in the direction of the meridian and take out by covering the point and taking out in 3 steps and once it is removed push the finger that is covering the point in the reinforcing direction.

9.3 Details about the directions for rotating and twisting the needle

Rotating and twisting needle manipulation, with your right hand on the right side of the patients body then moving the thumb backwards is reducing, forwards is reinforcing. If you are using reinforcing rotating and twisting needle manipulation with your right hand on the patients left side then move the thumb forwards, this is for reducing and backwards is reinforcing. On the Governing Vessel (GV) thumb forward with the right hand is reducing and on the Conception Vessel (CV) thumb backward is reducing.

9.4 Details about combining rotating and twisting with lifting and thrusting for inserting and removing the needle

We cannot use just rotating and twisting needle manipulation because the muscle tissue will wrap around the needle and it will get stuck. So we combine rotating and twisting needle manipulation with lifting and thrusting. This enables us to do needle manipulation without the needle getting stuck.

The reducing method of needle manipulation using lifting and thrusting combined with rotating and twisting on the right side of the patients body using our right hand would be 3 steps insertion with 3 times thumb backwards against the direction of the meridian and remove by stretching the skin and taking out in 1 step with thumb moving backward. The reinforcing rotating and twisting needle manipulation using the reinforcing lifting and thrusting method on the right side with the right hand would be 1 step in with the thumb moving forward and covering the hole and 3 steps out moving the thumb forward and pressing the hole in the direction of the meridian.

9.5 Details for combining rotating & twisting with lifting & thrusting of the needle once inserted inserted

The needle has been inserted as described above and the reducing needle manipulation method is now going to be applied. Whilst the needle is moved up and down and turned it remains under the skin.

For reducing manipulation it is a 3 step insertion of the needle and a 1 step pulling the needle back up motion whilst simultaneously rotating the thumb backwards with our right hand on the right side of the patient.

6 x 2 for *yin* reducing and for extreme excess 2 x 6 x 2. For beginners doing reducing manipulation start of with 3 x 3mm in and a 1 x 9mm out. Advanced would be 3 x 1mm in and 1 x 3mm out. The reinforcing needle manipulation method is 1 step insertion of the needle and 3 step pulling the needle back up motion whilst simultaneously rotating the thumb forwards with our right hand on right side of the patient.

9 x 3 for *yang* reinforcing and for extreme deficiency 3 x 9 x 3. For beginners doing reinforcing manipulation start of with 1 x 9mm in and 3 x 3mm out. Advanced would be 1 x 3mm in, and 3 x 1mm out.

Chapter 10

Traditional Chinese Diagnosis

Traditional Chinese Diagnosis is comprised of many different aspects which when combined reveal to the practitioner the underlying causes of the patients condition. These are: the Five Methods, Five Elements, Eight Principles, Six Stagnancies and face, tongue, eye and pulse diagnosis.

10.1 The Five Methods

The five methods are :

- *Interrogation:* asking – medical history, symptoms, diet and lifestyle
- *Inspection:* looking – observations of the face, tongue and lips
- *Olfaction:* smelling – alcohol on breath?
- *Auscultation:* listening – press ear to upper back to hear if phlegm is stuck on the back
- *Palpation:* pulse & body – taking the pulse on the radial artery, finger pressure on the body and spine to locate hard masses and inflammation

10.2 Basic patient consultation form

Patient Consultation Form

The Clinic of The College of Chinese Medicine

Name : Age: Date :

How did you hear about me/this clinic:

Address :

Tel : Occupation :

GP Name : GP Address :

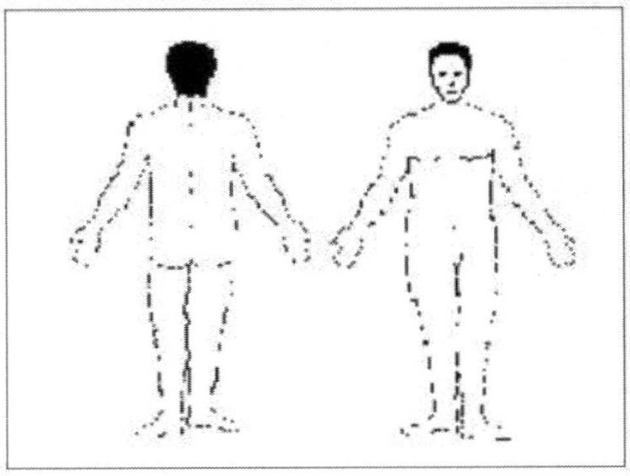

Medication:

Hepititus:

Illnesses :

Injuries :

Operations :

Family Medical History :

Food Allergies or Intolerances :

Breakfast :

Lunch :

Dinner :

Soft drinks :

Tea :

Coffee :

Alcohol :

Tabacco :

10.3 The Five Elements

Element	**Wood**	**Fire**	**Earth**	**Metal**	**Water**
***yin* Organ**	Liver BLOOD	Heart/PC BLOOD	Spleen BLOOD	Lungs CHI	Kidneys ESSENCE
***yang* Organ**	Gall Bladder	S.I. / T.W.	Stomach	L.I.	Bladder
Orifice	Eyes	Tongue	Lips (Sp.) Mouth (St.)	Nose	Ears
Sense	Sight	Speech	Taste	Smell	Hearing
Nourishes	Tendons Nerves Nails	Blood Vessels, Pulse, Complexion	Muscles	Skin, Body & Head Hair	Bones, Teeth, Genitals, Pubic Hair BACK L4, L5
EPA	Wind	Heat TIREDNESS INFLAMMATION	Damp HEAVYNESS	Dryness	Cold CONTRACTION PAIN
Colour	Green	Red	Yellow	White	Black
Excessive Emotion	Anger, Depression, Irritability	Hate, Impatience, Restlessness	Worry, Anxiety, Obsession	Sadness/Grief, Depression, Anguish	Fear
Deficient Emotion	Kindness	Love, Joy	Sympathy Fairness	Courage Uprightness	Gentlessness Contemplation
Sound	Shout	Laugh	Sing	Weeping	Groan
Secretion	Tears	Sweat	Saliva	Mucus	GU Fluids
Season	Spring	Summer	Late Summer	Autumn	Winter
Time of Life	Birth	Adolescence	Mid-Age	Old Age	Death
Flavour	Sour	Bitter	Sweet	Pungent	Salty
Meat	Chicken	Sheep	Beef	Horse	Pork
Excess	Walking	Watching	Sitting	Lying	Standing
Damages	Tendons	Blood	Muscles	Energy	Bones
Abstract	Hun/ Ethereal Soul	Shen/ Spirit	Yi/ Intellect	Po/ Animal Spirit	Zhi/ Willpower

魂 WEN 神

10.4 Explanation of the Five Elements

10.4.1 Wood – *Wind-Liver-Tendons-Eyes-Spring*

Symptoms

Clinical manifestations include : Itching, eye pain, abdominal pain, hiccups, burping, flatulence, tendon problems, a fibrillation running along just under the skin surface. A fibrillation is the rapid, irregular, and unsynchronised contraction of muscle fibers.

Diagnosis

- Eyes : blue sclera
- Lips : blue around lips
- Pulse : Floating

Commentary

If the body's *qi* and blood are weakened, Exogenous Wind, usually in combination with Cold, Damp or both, can penetrate the defensive level and affect the body.
Stagnation of wind causes irritation, tension, soreness and pain of the skin, mucous membranes of respiratory and digestive tracts, muscles and joints, with the pain moving from joint to joint and limitation of movement.

A floating pulse reflects the exterior of the body where the struggle between the *wei qi* (defensive energy) and the pathogenic factor takes place. Wind is reflected as tension in the pulse.

TCM attributes the following symptoms to Wind pathology: stiff neck, headache along the GB meridian, spasm and contraction of the tendons and ligaments, irritation, migratory body aches, stagnation of Liver and GB bile secretions into intestines resulting in flatulence, hypochondriac distention and constipation. Wind can also produce tics and itchy feeling under skin.

Acupuncture

Table 10.1: Points used Reducing

Point	Actions
BL.10, GB.20, GB.14	Clears Wind from the head
GV.14	Clears Wind from the head and UBC
GV.16	Clears Wind and tension from the head
BL.11, BL.12	Cup to clear Exogenous Wind from Lungs
SI.3 + BL.62	Trigram Point for stiff upper back and neck
LI.11	Clears Lung Heat, induces sweating and relieves itching
LIV.3	Balances Liver to resist Wind
GB.38	Clears Wind Headache from GB meridian
GB.14	Clears Wind-Headache from forehead, and strengthens body's resistance to wind in upper back

Herbs to alleviate symptoms of wind

- **Jing Jie** : exogenous wind-cold
- **Fang Feng** : exogenous wind-cold
- **Ju Hua** : exogenous wind-heat
- **Bo He** : exogenous wind-heat
- **Chai Hu** : stagnation of Liver and exogenous wind
- **Hai Feng Teng** : wind-damp good for joints
- **Bai Ji Li** : wind heat
- **Bai Zhi** : wind-cold-damp
- **Chuan Xiong** : deficiency blood with wind
- **Yu Jin** : stagnation of blood and *qi* of Liver

10.4.2 Fire – *Heat-Heart-Blood-Tongue-Summer*

Symptoms

Clinical manifestations include : tiredness, inflammation, redness, drying of fluids, constipation.

Diagnosis

- Eyes : red sclera, red under eyelids
- Lips : dark red, cracked.
- Tongue : red taste buds, red surface colour, yellow fur, dry.
- Nose : thick yellow nasal phlegm, red nostrils and nosebleeds.
- Other : hot palms and soles, red cheeks, hard and dark faeces, blood in stool, blood in urine and anal bleeding.
- Pulse : rapid, forceful, full, hesitant

Commentary

The battle of the immune system fighting against an external pathogen generates heat which then becomes pathogenic. Exaggerated amounts of hot foods such as chocolate, coffee, red meat, chilli sauce, garlic, onion, cinnamon and alcohol will warm the interior and the blood.

Aggressive emotions such as anger, hatred, bitterness and frustration can cause Liver *qi* & blood stagnation. As a result, the blood heats up the Liver and the whole body. You will see an urgent, rapid and forceful pulse: flushed face, red tongue, red sclera, reddened flushed lips and face all indicate heat. The tongue tip will be particularly red and tip fissures may appear. Surface fur will become yellow in a strong patient and whitish yellow in weak patients. Heat will lead to sweating and dehydration (if no sweating: diaphoretics are used. If profuse sweating in deficient patients: astringent tonics are used). Heat is associated with general symptoms of feeling tired, sore throat and headache, whole body ache, nausea, diarrhoea or constipation.

Acupuncture

Table 10.2: Points used Reducing

Point	**Actions**
Jing Wells	(SB) Clears heat.
TW.5, GB.41	Trigram points to clear heat and temperature.
LIV.3	Releases Liver meridian.
LU.7	Clears heat from the Lungs and throat.
ST.40	Clears Phlegm.
LI.11	Clears Lung and surface heat.
LU.10	Clears Lung skin and throat heat.
H.7	Clears Heart heat and calms mind.
GV.14	Clears UBC heat.
LI.4	Clears heat from the head.

Table 10.3: Points used Reinforcing

Point	**Actions**
KID.3, BL.23	Supports Kidney *yin* and body fluids.

Herbs to relieve symptoms of heat

- **Huang Qin** : clears heat fire, especially from the UBC. It also clears heat and dries dampness. It is good for Stomach and intestinal disorders with diarrhoea, dysentery, etc. Huang Qin brings down Liver *yang* Rising, therefore it treats headaches, flushed face, irritability, red eyes, and bitter taste in the mouth. Eliminates toxins for lesions, sores, ulcerations

- **Qing Dai** : clears heat, resolves toxicity, cools the blood, and reduces swellings. Used internally for Lung heat cough, childhood convulsions, high fever, and reckless blood that causes bleeding (like epistaxis & hemoptysis)
- **Bai Mao Gen** : clears heat in the blood, heat and pain in the urinary tract and heat in the Lungs. The herb Bai Mao Gen clears heat and promotes urination. It effectively treats dysuria with edema, urinary dysfunction such as painful, heat on urination. It clears Stomach and Lung heat. For cases of nausea, belching and thirst due to heat in the Stomach, cough and wheezing due to heat in the Lungs
- **Zhu Ru** : clears heat, dissolves phlegm. It is used for Lung heat cough with thick yellow phlegm, chest constriction, and hemoptysis. Treats conditions of Gall Bladder fire with phlegm, such as: irritability, insomnia, nausea, hypchondriac distension, bitter taste in mouth with profuse phlegm. It clears heat and stops vomiting, for example, Stomach heat vomiting of bitter and sour material, halitosis, yellow greasy tongue
- **Mu Dan Pi** : clears heat, cools the blood, treats heat entering the blood aspect, for irregular and profuse menstrual bleeding, epistaxis, hematemesis, hemoptysis. It cools false heat from *yin* deficiency

10.4.3 Earth – *Damp-Spleen-Muscles-Lips-Late Summer*

Symptoms

Clinical manifestations include : oedema, runny nose, thin sputum, heaviness, phlegm, puffiness, aching muscles, fatigue, heavy head, Irritable Bowel Syndrome.

Diagnosis

- Tongue : swollen body, moist fur, moist surface.
- Other : excess clear thin saliva, puffy cheeks.
- Pulse : soft, slippery.

Commentary

An urgent and difficult urination and a feeling of heaviness, which is typical of dampness, is caused because dampness is heavy. It obstructs the Water passages of the Lower Burner and interferes with the Bladder function of *qi* transformation. The pulse is softer and on a moderate layer the tongue will be moist with floating white fur. Dampness stagnates the Spleen function and its presence within the muscles produces lack of strength and aching muscle as the Spleen is the controller of absorption and the gastro intestinal tract and there appears sudden abdominal colic and the urge to defecate, tenesmus, diarrhoea nausea.

Herbs to treat symptoms of damp

- **Bai Zhu** : tonifies the Spleen and boosts the *qi*: It is used to treat diarrhoea, fatigue, lack of appetite, vomiting. It tonifies the Spleen and dries dampness:

Used for accumulation of damp due to improper transformation and transportation and difficulty raising the nutritive *qi* from the Spleen to the Lungs. It treats edema due to Spleen deficiency which is caused by accumulation of fluids. Also treats damp and excess fluid in the body.

- **Fu Ling** : promotes urination and drains dampness. It boosts the Spleen and harmonizes the middle burner: Used for diarrhoea, and upper abdominal bloating. It aids the Spleen to transform phlegm.
- **Yi Yi Ren** : powerful diuretic to relieve water retention in lower legs.
- **Hai Feng Teng** : dispels wind damp, frees the meridians: It is used for wind cold damp with painful and stiff joints, cramps and spasms of muscles and tendons, lower back and knee soreness and pain.
- **Zhi Ke** : moves stagnant *qi* and reduces distension: It treats food stagnation which causes constipation or diarrhoea, abdominal distension and pressure, indigestion, epigastric pain, flatulence. It has a down bearing effect, and moves the bowels: it is used for constipation and abdominal pain.
- **Gan Cao** : stops diarrhoea, regulates Spleen and detoxifies bacteria.

Acupuncture

Table 10.4: Points used Reinforcing

Point	Actions
SP.6, SP.9, SP.10	strengthens Spleen *qi* to clear damp and strengthen absorption
BL20	strengthens Spleen *yang* to clear off damp and activate digestion
BL23	strengthens Kidney to promote diuresis and expel water retention

10.4.4 Metal – *Dryness Lungs Skin Nose - Autumn*

Symptoms

Clinical manifestations include : dry skin, constipation, dandruff, bleeding, heat.

Diagnosis

- Tongue : dry fur.
- Lips : cracked lips.
- Other : dark concentrated urine, hard and dark faeces.

Commentary

Dryness is due to depleted body fluid and essence. It mainly affects the Lung and produces symptoms of dry cough, nose and throat: pulse becomes lacking in volume: tongue develops peeled patches without fur (bald). It can lead to bleeding lung nose and Stomach. Dryness appears on clear days and is especially pronounced during the autumn and winter.

Herbs to treat symptoms of dryness

- **Zhi Mu** : clears Lung and Stomach heat, and drains fire in the *qi* level: It treats high fever, thirst, irritability, rapid pulse. This herb nourishes Lung and Kidney *yin*, and moistens dryness: It is used for *yin* deficiency five centre heat, night sweats, , low grade fever, irritability, restlessness, bleeding gums, spermatorrhea, nocturnal emissions.
- **Tian Men Dong** : promotes the production of fluids and essence in Lung and Kidney, tonifies *yin* and clears heat: It is commonly used for *yin* deficiency heat with dry mouth, dry cough with thick or bloody sputum that is hard to expectorate. It moistens the Intestines, tonifies the Kidney and increases fluids: Use for constipation from dryness in the Intestines, or *yin* deficiency. Also for sore, dry throat and Lung heat.
- **Shi Hu** : clears heat and tonifies *yin* and generates body fluids. It treats severe thirst, dry mouth and fever from *yin* deficiency. It tonifies Stomach *yin* and is used for Stomach ache and diabetes. All these symptoms appear in the aftermath of a febrile disorder (EPA causing heat) and exhaustion of body fluid.
- **Nu Zhen Zi** : nourishes Liver and Kidney essence and clears *yin* deficiency heat. It relieves night sweats, mood swings, hot flush, tidal fever and irritability.
- **Xuan Shen** : nourishes the *yin*, increases fluids and clears deficient and latent heat. It treats constipation, irritability, sore throat, dry mouth,*yin* deficiency heat and reduces swelling of the glands, used in many anti EPA formulas.

Acupuncture

Table 10.5: Points used Reinforcing

Point	Actions
KID.3, SP.6	Promote fluids when they are deficient.
LU.1, KID.27	Moistens and nourishes the Lung

Table 10.6: Points used Reducing

Point	Actions
TW.5, GB.41	Clear heat

10.4.5 Water – *Cold - Kidneys - Bones Ears - Winter*

Symptoms

Clinical manifestations include : slowing down of organ metabolism, diarrhoea, pale face and tongue, lower back and lumbar soreness, infertility, impotence, erectile dysfunction, weak knees: weak, cold, slow.

Diagnosis

- Tongue : White fur.
- Nose : thin clear nasal phlegm
- Pule : slow.
- Other : cold body and extremities

Commentary

The feeling of cold, slow pulse, aggravation from cold liquids and preference for warm liquids is derived from the cold that impairs the *yang* of the Stomach and Spleen, as well as Kidney, which prevents the food essences from reaching the body.

The vomiting and the pain derive from exterior cold that blocks the Stomach and prevents Stomach *qi* from descending. External invasion of cold in the joints and the muscle tissue around the joint is cold painful obstruction.

Slower and deeper pulse, reduced circulatory power and functional capabilities and capacity of the body organs (especially Spleen, Kidney and Heart) to perform their usual duties, which may result in water retention and cold hands/feet and body aches.

Herbs to treat symptoms of cold

- **Hai Feng Teng** : warms Stomach and expels wind cold
- **Du Zhong** : tonifies the *yang* & strengthens and warms the Liver and the Kidney: It is used for tendons, ligaments and bone problems, weakness and pain of the lower back and knees, fatigue
- **Gui Zhi** : It is a powerful diaphoretic. It is used for weak circulation of the *yang* caused by cold stagnation. It warms the meridians and promotes circulation of *qi* and blood: clears wind damp cold in the joints, lower extremities and shoulders. It is also used for dysmenorrhea from cold blocking the blood flow

Acupuncture

Table 10.7: Points used Reinforcing

Point	Actions
KID.3, SP.6, SP.9, SP.10	For body resistance and stamina.
KID.12, CV.3, CV.4	For body resistance and stamina.
BL.15, BL.17, BL.20, BL.23	HE, GV, SP and KID Shu warms circulation and body
ST.36	Regulates food energy, nourishes and builds blood, improves nutrient absorption.

10.5 Explanation of The Eight Principles

- *yin* and *yang*
- External and Internal
- Excessiveness and deficiency
- Heat and Cold

10.5.1 *yin* and *yang*

yang refers to the functional strength or weakness of the internal organs. *yang* conditions have hyper-activity, increased metabolism, heat, excessiveness. External conditions are also referred to as *yang*.

yin refers to the excess or deficiency of essence of the internal organs and their substance. *yin* conditions have hypo-activity, decreased metabolism, debility, depression, cold, deficiency. Internal conditions are also referred to as *yin*.

10.5.2 External and Internal

External refers to the location of the disease on the surface of the body. Problems of the musculo-skeletal system fall under external diseases, especially when they are the principal problem and not just a symptom of internal disease.

External also means disease caused by Exogenous Pathogenic Factors. The body will become ill if attacked by excessive heat, cold, wind, dampness, dryness or by viruses, bacteria or toxic substances. The patient might have runny nose, cough and muscle aches. All these symptoms indicate that the exterior of the body is being attacked by Exogenous Pathogenic Factors.

Internal refers to chronic conditions where the disease has penetrated to the interior of the body and is residing in the internal organs. Internal also means disease caused by Endogenous Pathogenic Factors.

10.5.3 Excessiveness and deficiency

A patient with a deficiency condition could have deficiency of functional power of the internal organs, or deficiency of blood, or energy, or essence or body resistance. deficiency conditions can be caused by prolonged illness, incorrect lifestyle or due to weak inherited energy.

Excessiveness conditions have hyper-functioning in the interior which indicates an excessiveness of pathogenic factors and relative strength of the body resistance.
Excessiveness conditions are usually observed in the initial stage of a disease.
For excessive conditions we use reducing methods of treatment.
For deficiency conditions we use reinforcing methods of treatment.

10.5.4 Heat and Cold

A person with heat symptoms could have fever, red face, thirst, constipation, red tongue substance with yellow fur and a rapid pulse.

A person with cold symptoms could have weak functioning of the internal organs, an aversion to cold, fondness for warmth, diarrhoea, pale face and pale tongue substance with white fur, and thin, slow pulse.

yang and *yin* Heat

Heat can be split into two types - excessiveness *yang* heat and deficiency *yin* heat.

Excessiveness *yang* heat is usually a response to an attack from pathogenic factors and there is inflammation or temperature and it is treated by clearing heat.

deficiency *yin* heat feels like a hot flush and is caused by deficiency of essence. The patient with deficiency heat might also experience heat sensation in the palms, spontaneous or night sweating and general lethargy. This type of heat is treated by replenishing essence.

Excessiveness heat can lead to deficiency heat because Heat consumes essence. An example of this is when the flu becomes ME.

yang heat, or excessive type heat, is often caused by the immune system rising up to fight an EPA, the battle between the immune system and the EPA creates heat, although the EPA was originally the main problem, the excessive heat generated by the battle also becomes a problem, it causes inflammation, pain, and swelling and it dries the blood and burns off body fluid leading to dehydration, thirstiness and constipation, a red complexion, fever, thirst with preference for cold drinks, scanty and deep yellow coloured urine, red tongue substance, yellow tongue fur and a dry tongue coating, with full and forceful rapid pulse.

Herbs for *yang* heat

- **Huang Qin** : clears temperature and heat
- **Qing Dai** : clears temperature and heat, good for skin and low grade blood heat
- **Zhi Zhi** : clears excess Liver heat, relieves irritability and inflammation
- **Chi Shao Yao** : clears heat in the blood, and LBC
- **Mu Dan Pi** : clears heat in the blood, particularly good for EPA blood heat.
- **Da Huang** : cools intestines and purges bacteria and virus

Acupuncture for *yang* heat

Table 10.8: Points used for *yang* Heat

Direction	Points
Reinforce	K3, BL23.
Reduce	TW.5, GB.41, GB.34, LV.3, LU.10, LU.7, ST.36, LI.4, LI.11, LU.11 (SB), LI.1 (SB), BL.60, apex of ears (SB), GV & HT at lower border of T3, T4, T5, T6, T9, T10.

Herbs for *yin* heat

The manifestations of *yin* deficiency heat are hot flushes, warm palms and soles of the feet, heat sensation in the chest and face with red patches on the cheek bones, thirst with fast but fine or small and weak pulse. A reddish tongue substance without fur, lassitude and fatigue, hot flushes and night sweats.

- **Bai He** : replenish essence, moistens for dry cough, latent EPA
- **Sheng Di Huang** : nourishes essence blood and *qi*
- **Nu Zhen Zi** : nourishes Liver and Kidney *yin* to cool
- **Sang Shen** : nourishes essence to cool
- **Shi Hu** : nourishes kidney and lung essence eases dry throat/lung
- **Xuan Shen** : nourishes essence and clears *yin* deficient heat and sore throat
- **Lu Gen** : nourishes essence to clear deficient *yin* heat cools Lungs
- **Fu Xiao Mai** : stops sweating and replenishes *yin* to prevent further depletion
- **Mu Dan Pi** : clears blood heat
- **Shan Zhu Yu** : helps to maintain and retain the body fluid
- **Zhu Ru** : clears heat
- **Di Gu Pi** : replenishes essence for deficient *yin* heat and counters hot flushes
- **Tian Men Dong** : replenishes essence to cool temperature
- **Zhi Mu** : replenishes essence
- **Tao Ren** : replenishes essence to moisturise, thin and move blood
- **Bai Shao Yao** : nourishes essence of the blood

Acupuncture points for *yin* heat

Table 10.9: Points used for *yin* Heat

Direction	Points
Reinforce	KID.3, BL.23, BL.24, BL.52, SP.6, SP.9, SP.10, LU.1, KID.27.
Reduce	LI.4, LI.11, BL.60, TW.5, GB.41

10.6 TCM Syndromes

At the College of Chinese Medicine we primarily practice diagnostic TCM. In simple terms this means we find different things through our diagnosis and then we treat what we find. So if we find that a part of the body is dry we moisten that part of the body. If in another part of the body we find heat then we clear heat from that area. If there is damp we clear damp, if wind we clear wind. If weakness is found we reinforce, if excessiveness we reduce etc.

If we find a combination such as blood deficiency leading to invasion of wind and cold then we replenish the blood and clear the wind and cold.

However there is another approach which is called identifying syndromes. This method is different from the diagnosis method but can be used in conjunction with it.

When using the syndrome method the emphasis is to try and fit the patient into an already existing category (These categories or syndromes are listed below).

Patients very rarely fit neatly into one syndrome and often have a combination of syndromes mixed together. However learning the standard syndromes is an important part of Traditional Chinese Medicine.

10.6.1 Lung Syndromes

qi deficiency of the Lung

Feeble coughing, shortness of breath, clear and thin sputum, feeble breathing, speaking in a low voice, spontaneous sweating, pale and lustreless complexion, lassitude, pale tongue proper with thin white coating, deficient and weak pulse.

The lung dominates *qi* and controls respiration, so *qi* deficiency causes a weak cough, shortness of breath, and feeble breathing. *qi* deficiency of the lung also leads to a failure of lung *qi* to descend causing an accumulation of body fluid with resulting phlegm. There are also symptoms of cough with thin sputum, spontaneous sweating, pale and lusterless complexion, lassitude, pale tongue proper with white and thin tongue coating, and deficiency type pulse.

yin deficiency of the Lung

Dry cough without sputum or with a little sticky sputum, dryness of the mouth and throat, hoarseness of voice, emaciation, dry red tongue proper, thready and weak pulse. If *yin* deficiency leads to a preponderance of fire, there may be cough with bloody sputum, tidal fever, night sweating, malar flush, red tongue proper, and a thready rapid pulse.

The symptoms of lung *yin* deficiency are actually the manifestations of an insufficiency of lung *yin* fluid, i.e., dry cough without sputum, or cough with a little sticky sputum, dryness of the mouth and throat, hoarseness of voice, emaciation, dry red tongue proper, thready and forceless pulse. If *yin* deficiency fails to restrict *yang*, then deficiency fire is formed and flares up to damage the vessels of the lung, producing the symptoms of tidal fever, night sweating, malar flush, hematemesis, red tongue proper, thready rapid pulse.

Lung wind-cold retention syndromes

Cough, asthma, thin white sputum, absence of thirst, nasal obstruction, runny nose, chills and fever, no sweating, pain of the head and body, thin white tongue coating, superficial and tense pulse.

Exogenous pathogenic wind-cold obstructing the lung leads to the dysfunction of lung *qi* spreading and descending, causing symptoms of cough with thin white sputum. The lung opens into the nose, which is then also troubled by nasal obstruction or discharge. The lung dominates the skin and hair, when exogenous pathogenic wind and cold invade the lung leading to the dysfunction of defensive (wei) *qi*. The symptoms are an aversion to cold, fever, pain of the head and body, absence of sweat, thin white tongue coating.

Lung wind-heat invasion syndromes

Cough with yellowish and thick sputum, thirst, sore throat, headache, fever, aversion to wind, yellowish and thin tongue coating, floating and rapid pulse.

The lung is attacked by exogenous pathogenic wind-heat, so the spreading and descending functions are affected, causing cough with a yellowish and thick sputum. Pathogenic heat consumes the body fluid, causing thirst. Wind and heat flow upward to cause a sore throat. Headache, fever, aversion to wind, yellowish and thin tongue coating, floating and rapid pulse are signs indicating wind-heat invasion of the defensive (wei) *qi* of the body surface.

Lung phlegm damp obstruction syndromes

Cough with excessive and white sticky sputum, expectoration, stuffiness of the chest, asthma, white sticky tongue coating, slippery pulse.

Pathogenic phlegm damp obstructing the lung leads to the impairment of *qi* circulation causing the above symptoms. White sticky tongue coating and slippery pulse are signs of pathogenic phlegm damp.

A long-standing obstruction of phlegm damp in the lung will change into heat, blocking *qi* circulation and manifesting as asthmatic cough, stuffiness of the chest, etc. In addition, other symptoms may occur, such as cough with yellowish, sticky and thick sputum, or cough with bloody and pus in sputum. Fever, thirst, yellowish urine, constipation, red tongue proper with yellow sticky coating, and slippery pulse, are signs of heat syndromes.

10.6.2 Large Intestine Syndromes

Large Intestine damp-heat syndromes

Abdominal pain, dysentery or stool containing blood and pus, tenesmus, burning sensations of the anus, scanty and yellowish urine, yellow and sticky tongue coating, wiry, slippery and rapid pulse.

The retention of damp-heat in the Large Intestine causes a dysfunction of *qi* activity with resulting abdominal pain and tenesmus. Damp-heat injures the *qi* and blood of the intestinal tract, so dysentery, or bloody and purulent stool occur. Burning sensation of the anus is a characteristic manifestation of 'downward pouring of damp-heat into the Large Intestine. Scanty yellowish urine, yellow sticky tongue coating, and wiry, slippery, and rapid pulse are signs of internal retention of damp-heat.

Large Intestine fluid exhaustion

Constipation, difficult defecation of dry stools, dryness of the mouth and throat, red tongue proper with a yellow dry coating and a thready or rough pulse are all signs of fluid consumption.

10.6.3 Spleen Syndromes

Spleen failure to carry out transportation and transformation

Anorexia, abdominal distension after meals, lassitude, sallow complexion, feeble breathing, loose stool, pale tongue proper with white thin coating, retarded and weak pulse.

Spleen deficiency causes a failure of transportation and transformation, and insufficiency of *qi* and blood, so the above symptoms appear.

Sinking of Spleen *qi* syndromes

Prolapse of the uterus, gastroptosis (collapse of the Stomach), nephroptosis (collapse of the Kidneys), chronic diarrhoea, feeble breathing, yellowish complexion, pale tongue proper with white coating, and deficiency type pulse.

Spleen *qi* should ascend, however, Spleen deficiency causes the *qi* to sink. If the Spleen *qi* is too weak to elevate the internal organs, then the prolapse of internal organs and symptoms showing Spleen *qi* insufficiency occur.

Spleen blood control failure syndromes

Excessive menstruation, uterine bleeding, anal bleeding, bloody urine, purpura, pale complexion, lassitude, pale tongue proper, and a thready weak pulse.

The Spleen controls blood. If it is unable to carry out this function, then bleeding occurs, plus the bleeding symptoms mentioned above. Bleeding affects the function of transportation of *qi* and blood, resulting in a pale complexion, lassitude, a pale tongue proper, and a thready weak pulse, which are signs of *qi* and blood deficiency.

Pathogenic damp invasion of the Spleen syndromes

Distension and fullness of the epigastrium and abdomen, anorexia, stickiness in the mouth, heaviness of the head, absence of thirst, swelling of the face, eyes, and four extremities, loose stool, dysuria, and leucorrhoea, white and sticky tongue coating, and soft thready pulse.

The Spleen is adverse to dampness, therefore excessive dampness is liable to affect Spleen *yang* leading to a dysfunction of transportation and transformation, resulting in the symptoms of distension and fullness of the epigastrium and abdomen, and anorexia.

Pathogenic damp, which is sticky and stagnant in nature, easily blocks the flow of *yang* qi, causing a sensation of heaviness of the head. If dampness and fluid pour into the skin and muscles, swelling of the face, eyes, and extremities occurs. If the Spleen fails to remove the damp, the stool becomes loose and the urine abnormal. A white and sticky tongue coating and a soft thready pulse are signs of excessive pathogenic damp.

Spleen *yang* deficiency syndromes

Dull pain of the epigastrium and abdomen ameliorated by warmth, chills with cold extremities, poor appetite, loose stool, pale tongue proper with white coating, and deep, slow pulse.

Spleen *yang* deficiency causes the stagnation of cold in the middle jiao, obstructing the functions of *qi*. Warmth can remove the obstruction, so the pain of the epigastrium and abdomen is ameliorated. deficiency of Spleen *yang* leads to a dysfunction of transportation and transformation, thus the failure of Spleen *yang* to warm the body surface and extremities, and the occurrence of anorenxia, and loose stool. A pale tongue proper with a white coating and a deep slow pulse are signs of deficiency cold.

Spleen and Stomach damp heat syndromes

Yellow complexion, distension and fullness of the epigastrium and abdomen, nausea, vomiting, poor appetite, aversion to greasy food, heaviness of the body, yellowish urine, loose stool profuse and yellowish leucorrhoea, yellowish and sticky tongue coating, soft and rapid pulse.

Damp heat accumulates in the skin causing a yellow-orange complexion. It also blocks the middle jiao causing symptoms of distension and fullness of the epigastrium and abdomen, nausea, vomiting, anorexia, and aversion to greasy food. Excessive damp causes heaviness and tiredness of the body. Damp heat descending leads to profuse yellowish leucorrhoea. Deep yellow urine, loose stool, yellowish and sticky tongue coating, and soft pulse are signs of excessive damp heat.

10.6.4 Stomach Syndromes

Loss and deficiency of Stomach *yin*

Dryness of the mouth and throat, Stomach-ache and hunger without desire to eat, dry stool, red tongue proper with scanty fluid, thready and rapid pulse.

Insufficiency of Stomach *yin* makes the body fluid fail to support the upper organs, causing dryness of the mouth and throat. Insufficiency of Stomach fluid leads to the dysfunction of Stomach reception manifested by hunger without desire to eat. deficiency of Stomach *yin* also gives rise to the disturbances of deficiency fire, manifesting as Stomach pain. Insufficiency of Stomach *yin* causes dry stool. Red tongue proper with scanty fluid and a thready rapid pulse are signs of *yin* deficiency producing heat.

Preponderance of Stomach fire

Burning pain of the epigastric region, vomiting, nausea, acid regurgitation, constipation, thirst with preference for cold drinks, swelling, pain, ulceration and bleeding of the gums, hunger with excessive eating, foul breath, red tongue proper with yellow coating, slippery and rapid pulse.

Accumulation of heat in the Stomach leads to a dysfunction of *qi* activities resulting in a burning pain of the epigastrium. Preponderance of heat in the Stomach consumes the *yin* of the Stomach causing thirst with a preference for cold drinks. Since pathogenic fire accelerates food, there is hunger with excessive eating. Branches of the Stomach channel travel up to the gum, therefore when pathogenic Stomach heat flows upward, it causes swelling, pain, ulceration and bleeding of the gums. An accumulation of Stomach heat leading to a dysfunction of Stomach *qi* descent causes foul breath, vomiting, nausea, and acid regurgitation. A red tongue proper with yellow coating and a slippery rapid pulse are signs of Stomach heat.

Retention of food in the Stomach

Distension or pain in the epigastrium, foul belching, acid regurgitation, no desire to eat, vomiting, abnormal bowel movements, diarrhoea or constipation, thick sticky tongue coating and slippery pulse.

Retention of food in the Stomach blocks the *qi* activities of the MBC, so there is distension or pain in the epigastrium. Foul belching, acid regurgitation, no desire to eat, and vomiting are caused by a dysfunction of Stomach *qi* descent, which then causes the upward flow of turbid *qi*. Retention of food in the Stomach affects the transportation and transformation functions of the Spleen, producing abnormal bowel movements, i.e., diarrhoea or constipation. A thick sticky tongue coating and a slippery pulse are signs of food retention.

10.6.5 Heart Syndromes

Syndromes of Heart *qi* deficiency and Heart *yang* deficiency

Palpitation and shortness of breath aggravated by exertion, spontaneous sweating, thready and weak pulse, and regular pulse or irregular intermittent pulse, are the basic symptoms of Heart *qi* deficiency and Heart *yang* deficiency. If the above symptoms are accompanied with a pale and lustreless complexion, lassitude and a pale tongue proper with whitish coating, they are in the category of Heart *qi* deficiency. If they are complicated with chills, cold extremities, fullness of the chest, pallor, and a pale or dark purplish tongue proper, they are considered as syndromes of the Heart *yang* deficiency.

If Heart *qi* or Heart *yang* is insufficient, then the blood circulation is not promoted and shortness of breath aggravated by exertion will manifest. If Heart *yang* is inadequate to restrict Heart fluid, there will be spontaneous sweating. *qi* deficiency leads to blood deficiency and weakness of *yang qi*, so disorders of blood circulation will manifest as thready, weak, irregular or regular intermittent pulses. Heart *qi* deficiency, or the failure of *yang qi* and blood to nourish the tongue, face and body, causes a pale and lusterless complexion, pale tongue proper, and lassitude. Heart *yang* deficiency fails to warm the limbs, which causes chills and cold extremities. Failure of *yang qi* in the chest causes the improper circulation of *qi* and blood, manifesting a fullness in the chest and a dark purplish tongue proper.

Syndromes of Heart blood deficiency and Heart *yin* deficiency

Palpitation, insomnia, dream disturbed sleep and poor memory. If these symptoms are accompanied with a lustreless complexion, dizziness, pale tongue and lips, and a thready pulse, then these are Heart blood deficiency syndromes. If the symptoms are complicated with irritability, thirst, feverish sensation of the palms and soles, tidal fever, night sweating, dry red tongue proper, and a thready rapid pulse, then these are Heart *yin* deficiency syndromes.

The Heart dominates the blood and its vessels, so Heart blood deficiency and Heart *yin* deficiency both cause malnourishment of the head region, manifesting as malnourishment of the mind, producing symptoms of palpitation, poor memory, insomnia, and dreamed disturbed sleep; malnourishment of the facial region, producing symptoms of lustreless complexion, pale tongue and lips; malnourishment of the brain, manifesting as dizziness and a thready weak pulse. Heart *yin* insufficiency also leads to Heart *yang* preponderance and internal deficiency fire disturbances which cause irritability, feverish sensation of the palms, dry red tongue proper with scanty fluid, and thready rapid pulse.

Syndromes of Heart fire preponderance

Ulcers of the tongue and mouth, anxiety, insomnia, thirst, yellowish urine, a red tongue tip, and rapid pulse.

The Heart opens to the tongue. If there is a preponderance of Heart fire, it flares up to attack the tongue causing ulceration. If Heart fire causes internal disturbances, it first affects the mind, causing irritability and insomnia. A preponderance of Heart fire consumes the body fluids, causing thirst, red tongue tip, and rapid pulse.

Stagnation of Heart blood syndromes

Palpitation, paroxysmal pricking pain, or stuffy pain of the pericardiac region referring to the shoulder and arm of the left side, cyanosis of the lips and nails, cold extremities, spontaneous sweating, dark red tongue proper, or purplish tongue proper with petechiae, thready rugged pulse, or regular and irregular intermittent pulse.

Obstruction of Heart *yang* leads to unsmooth circulation of *qi* and blood, and the stagnation of blood in the vessels, causing palpitation and cardiac pain. The Small Intestine channel is exteriorly and interiorly related to the Heart channel, so the *qi* of the two channels affect each other, that is why cardiac pain is related to the shoulder and arm. The stagnation of Heart blood may also bring on cyanosis of the lips and nails, dark red tongue proper, or purplish tongue proper with petechiae, thready rugged pulse, or regular and irregular intermittent pulse. Heart blood stagnation blocks the *yang qi* from spreading over the body surface and the four extremities, so cold extremities and spontaneous sweating result.

Phlegm fire Heart-disturbing syndrome

Mental disorder, weeping and laughing without apparent reason, mania, redness of face, thirst, coarse breath, yellowish urine, yellow and sticky tongue coating, slippery, rapid, and forceful pulse.

Phlegm-fire disturbs the Heart mind and exhausts the body fluid, so the above symptoms and pulses appear.

10.6.6 Small Intestine Syndromes

Syndromes of excess heat in the small intestines: (1) Scanty yellowish urine; burning pain of the urethra, or hematuria; ulceration and pain of the mouth and tongue; a feverish sensation with irritability in the chest. (2) The Heart has an exterior and interior relation to the Small Intestine, so a preponderance of Heart fire will transmit to the Small Intestine resulting in excess heat syndromes of the Small Intestine.

10.6.7 Urinary Bladder Syndromes

Syndromes damp and heat in the urinary Bladder: (1) Frequency, urgency and pain of urination; dribbling urination; turbid urine of bloody and purulent urine; urine with stones; a yellow sticky tongue coating; and rapid pulse. (2) The accumulation of damp-heat in the urinary Bladder blocks *qi* activity, causing dribbling urination. A downward driving of damp-heat into the urinary Bladder brings about frequent, urgent and painful urination. Bloody or purulent urine is due to the injury of blood vessels by damp-heat.

10.6.8 Kidney Syndromes

Kidney *yang* deficiency syndromes

Chilliness, cold extremities, aching and weakness of the lumbar region and knee joints, impotence, premature ejaculation, excessive and thin leucorrhoea, infertility, profuse and clear urine or enuresis, pale tongue proper with white coating, keep, slow and forceless pulse.

The kidney stores essence which is the original source of reproduction, therefore kidney *yang* deficiency will influent the genital system. Symptoms seen in men are impotence and premature ejaculation, and in women excessive and clear leucorrhoea, and infertility. The kidney dominates the bones and is the site of primary *yang qi* convergence. Insufficient *yang qi* of the kidney fails to warm and nourish the body and extremities, causing chilliness, aching and weakness of the lumbar region and knee joints. The kidney dominates water metabolism, so kidney *yang* deficiency causes a dysfunction of urinary Bladder restriction, manifesting enuresis or profuse and clear urine. A pale tongue proper with white coating and a deep, slow and forceless pulse are signs of *yang* deficiency.

Kidney *yin* deficiency syndromes

Dizziness, vertigo, ringing in the ears, deafness, hair loss, loosening teeth, soreness and weakness of the lumbar region and knee joints, insomnia, poor memory, dryness of the throat, night sweating, feverish sensation of palms and soles, low fever, malar flush, red tongue proper and thready, rapid pulse.

yin deficiency produces internal heat, so symptoms such as low fever, malar flush, feverish sensation of the palms and soles, and night sweating occur. *yin* deficiency also leads to insufficiency of body fluid manifested by dryness of the throat. Consumption of kidney *yin* causes soreness and weakness of the lumbar region and knee joints, hair loss and loosening teeth. *yin* deficiency also causes the kidney to fail in its function of producing marrow, and with it filling out the brain. Manifestations are dizziness, vertigo, poor memory and insomnia. *yin* deficiency is unable to nourish the upper orifices, and is manifested by ringing in the ears and deafness. Red tongue proper, and thready and rapid pulse are also signs of *yin* deficiency.

Kidney *qi* deficiency syndromes

Shortness and weakness of breath, asthmatic breathing aggravated by exertion, perspiration, cold extremities, swelling of the face, pale tongue proper, deficiency type pulse, etc.

The kidney dominates the reception of *qi*, so its weakness causes the *qi* to lose its function of controlling reception. The symptoms of shortness and weakness of breathing result. Asthmatic breathing aggravated by exertion is due to the consumption of *qi*. The deficiency condition of the kidney brings on *yang* deficiency leading to the weakness of wei (defensive) *qi*, so symptoms of perspiration appear. Cold extremities are due to *yang qi* failing to reach and warm the four extremities. *yang* deficiency also has difficulty in promoting *qi* circulation and water metabolism, so there is swelling of the face. Pale tongue proper and deficiency pulse are also signs of kidney *qi* deficiency.

Kidney deficiency leading to excessive water

General oedema with greater severity in the lower extremities, abdominal distention, scanty urine, short breathing, cough and asthma with sputum gurgling in the throat, palpitations, asthma aggravated by exertion, chilliness and cold extremities, flabby tongue body with white coating and deep thready pulse.

The declining of kidney *yang* causes a dysfunction of the urinary Bladder *qi* activity, manifesting as scanty urination. General oedema is due to the water and fluid overflowing into the skin and muscles. Retention of water and fluid in the abdominal cavity gives rise to local distension. Excess water and fluid converts into phlegm, manifesting as cough and asthma with sputum gurgling in the throat. If water and fluid overflow upward they attack the Heart and lung causing symptoms of palpitation and shortness of breath. *yang* deficiency fails to warm and nourish the extremities, so it causes chilliness and cold extremities. Flabby tongue body, white tongue coating, deep and thready pulse are signs of *yang* deficiency causing an overflow of water and fluid.

Unconsolidated kidney *qi* syndromes

Frequent and clear urination, incontinence, dribbling of urine, nocturnal enuresis, involuntary seminal discharge without dreams, premature ejaculation, soreness and weakness of the lumbar region, wan complexion, pale tongue proper with white coating, thready and weak pulse.

The kidney stores essence, if kidney deficiency fails to consolidate the source of semen involuntary seminal discharge and premature ejaculation occur. Kidney deficiency causes the dysfunction of urinary Bladder restriction seen in the symptoms of frequent and clear urination, dribbling of urine, incontinence, and nocturnal enuresis. The waist is the house of the Kidneys, deficiency causes soreness and weakness of the lumbar region. Wan complexion, pale tongue proper with white coating, and thready weak pulse are signs of *yang* deficiency in the kidney.

10.6.9 Gall Bladder Syndromes

Phlegm disturbing the Gall Bladder: (1) Dizziness, vertigo, bitter taste in the mouth, nausea, vomiting, irritability, insomnia, fright, fullness of the chest, sighing, slippery and sticky tongue coating, wiry pulse. (2) The Gall Bladder channel travels up to the head and eyes, so dizziness and vertigo are caused by pathogenic phlegm disturbing the brain along the course of the Gall Bladder channel. Internal phlegm disturbances lead to a restlessness of Gall Bladder *qi* resulting in irritability, insomnia, and fright. Stagnation of Gall Bladder *qi* affects the free flow of *qi*, this causes fullness of the chest and sighing. Since bile streams upward, there is a bitter taste in the mouth. *qi* stagnation of the Gall Bladder also disturbs the Stomach *qi*'s descent leading to nausea and vomiting. A sticky and slippery tongue coating and wiry pulse are signs of phlegm obstruction.

10.6.10 Liver Syndromes

Syndromes of Liver blood insufficiency

Dizziness and vertigo, distending pain, redness of the eyes and face, anxiety and hot temper, dryness of the eyes, blurred vision, night blindness, numbness of the limbs, spasm of the tendons and muscles, scanty menstrual flow or amenorrhea, pale tongue proper, and thready pulse

Insufficiency of Liver blood brings about malnutrition of the head and eyes, and manifests as dizziness, dryness of the eyes, and blurred vision. Consumption of Liver blood causes malnourishment of the tendons, manifested by numbness of the limbs, and spasms of the tendons and muscles. The thrusting channel's "sea of blood" dries up due to an insufficiency of Liver blood, so scanty menstrual flow or amenorrhea appears. Blood insufficiency also causes a pale tongue proper and a thready pulse.

Liver fire flare up syndromes

Dizziness, distending pain redness of the eyes and flushed face, irritability and irascibility, dryness and bitter taste in the mouth, deafness, ringing in the ears, burning pain of the costal and hypochondriac regions, yellowish urine, constipation or vomiting blood (hematemesis), and nose bleeds (epistaxis), red tongue proper with yellow coating, wiry and rapid pulse.

Liver fire flares up to attack the head and eyes causing dizziness, distending pain, redness of the eyes and flushed face, bitter taste and dryness in the mouth, deafness and ringing in the ears. Fire injures the Liver causing a dysfunction of the *qi* flow and since the Liver is related to emotional activities, depression and anger can result. As the Liver channel passes through the costal and hypochondriac regions, it causes pain in these areas.

Liver fire exhausts the blood and injures the vessels, so hematemesis and epistaxis occur. Yellowish urine, constipation, yellow tongue coating, and a rapid pulse are also signs of excessive Liver fire.

Liver *qi* stagnation syndromes

Fullness of the chest, mental depression, sighing, distending pain of the chest and hypochondrium, irascibility, anorexia, belching, abnormal bowel movements, irregular menstruation, dysmenorrhea, premenstrual distending pain of the breasts, thin and white tongue coating, and wiry pulse.

Stagnation of Liver *qi* leads to the dysfunction of the Liver causing an unrestrained flow of *qi*, so symptoms such as mental depression, fullness of the chest, irascibility, and sighing appear. Liver *qi* can also flow transversely to attack the Stomach and Spleen, causing disorders of the ascending and descending Stomach and Spleen *qi*. Symptoms of belching, anorexia, and abnormal bowel movements result. The Liver stores blood, so Liver *qi* stagnation will certainly affect menstruation causing irregularity, dysmenorrhea, or pre-menstrual distending pain of the breasts. A wiry pulse is also caused by Liver *qi* stagnation.

Liver wind stirring syndromes

There are three conditions commonly seen in the clinic: (1) Extreme heat stirring up endogenous wind, manifesting as high fever, convulsion, neck rigidity, contracture of the four limbs, rigidity, red tongue proper and a wiry rapid pulse. (2) *yin* deficiency leading to *yang* preponderance, this extreme *yang* then turns into wind and manifests as sudden temporary loss of consciousness (syncope), convulsion, deviated mouth and eyes, tongue rigidity, hemiplegia, wiry, slippery, and forceful pulse. (3) Insufficiency of Liver blood causes the malnutrition of tendons and muscles, and produces wind, manifesting as numbness of the limbs, tremor of muscles or spasms of the extremities, tremor of the hands, pale tongue proper and a wiry and thready pulse.

The first condition is a group of excess syndromes caused by extreme heat producing wind, the wind and fire then stir each other. The second condition originates from a loss of Liver and kidney *yin* which leads to Liver *yang* preponderance and an upward flow of *qi* and blood. The root of this disease is therefore deficiency, but symptomatically the syndromes appear as excess type. The third condition is also a deficiency condition due to insufficiency of blood which leads to malnutrition of the tendons and muscles.

Stagnation of cold in the Liver channel syndromes

Distending pain of the lower abdomen, swelling and distension of the testes with a bearing down pain, pain and contracture of the scrotum referring to the lower abdomen, a white slippery tongue, and a wiry slow pulse.

The Liver channel curves around the external genitalia and passes through the lower abdominal region. Pathogenic cold is characterized by contraction and stagnation when it inhabits the Liver channel. This results in the stagnation of *qi* and blood and causes the above symptoms.

10.6.11 Syndromes Overview

Students of the College of Chinese Medicine should be familiar with all these syndromes and be able to incorporate this perspective into their treatments.

However it cannot be emphasised enough that the primary method for deciding which points and herbs to use in ones treatment is from the unique diagnosis of the patient that is acquired through the understanding of the pulse, tongue, eyes, lips and face in conjunction with the symptoms of that unique patient on that day.

Chapter 11

Face Diagnosis

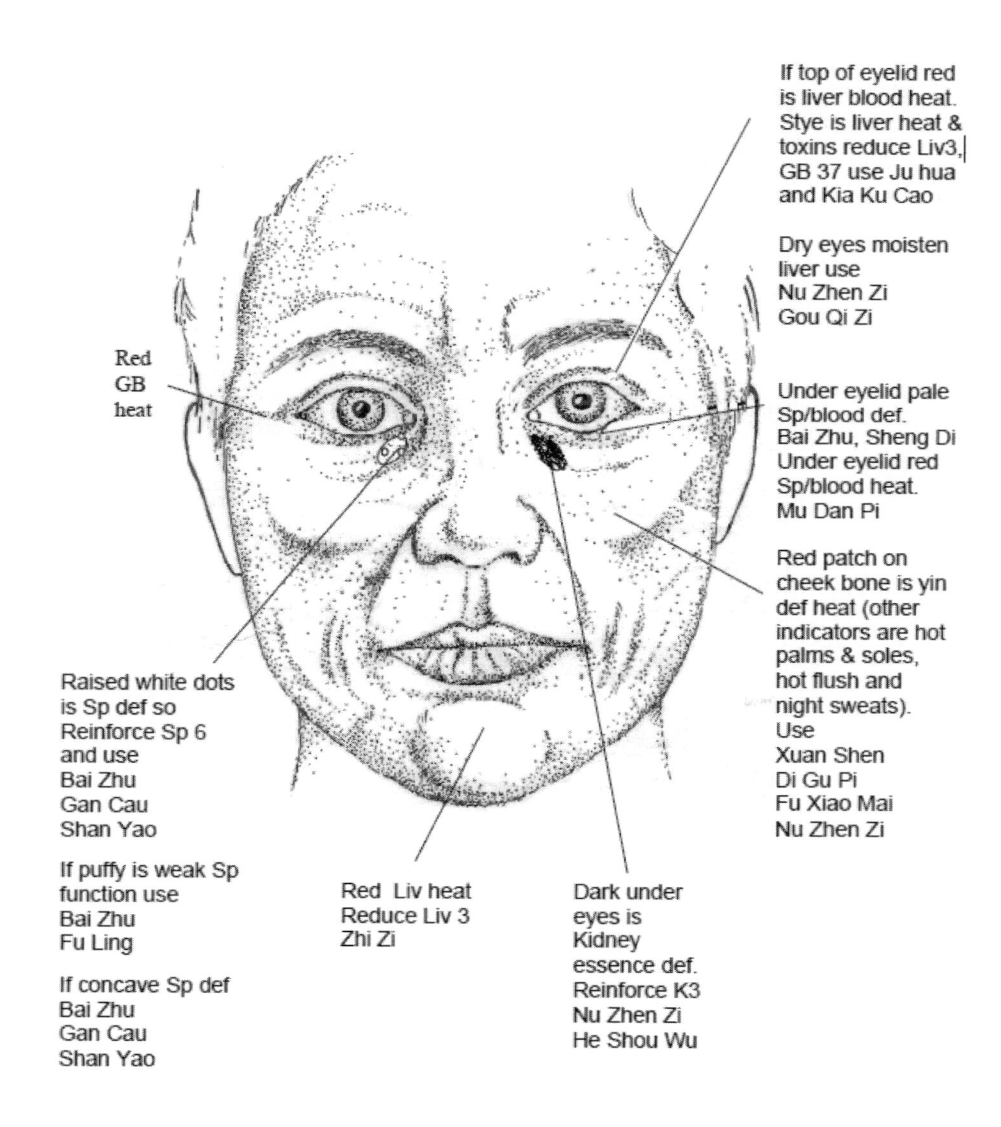

Whole face pale - deficiency blood or damp, if damp might also be a bit puffy. Whole face red - excess heat in body from fever or alcohol.

Chapter 12

Tongue Diagnosis

The tongue fur, the shape of the tongue, the colour of the tongue substance, the taste buds and condition of the surface of the tongue all indicate to us a great deal of information about the health of the body. By observing them the practitioner can form a diagnosis of the patients health. The patient should not eat for one hour prior to being diagnosed. The tongue is a muscle (Spleen) with a tendon (Liver) in it so its condition reflects the general condition of the muscles and tendons. The tongue is also like a large blood vessel (Spleen, Heart, Liver) so we can see the composition of the blood and the condition of the blood vessels.

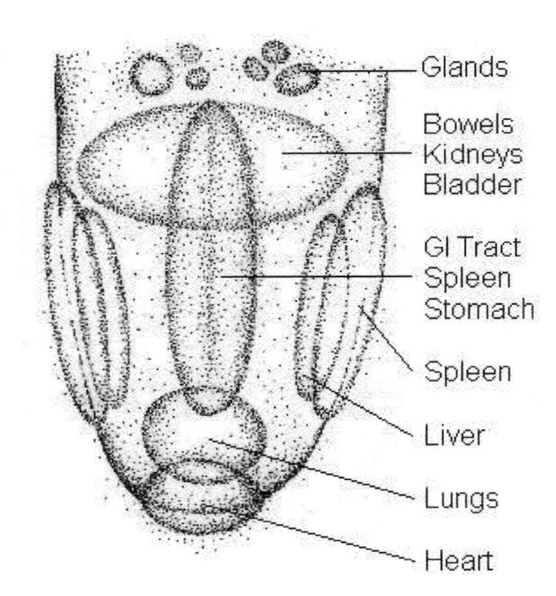

- The top third of the tongue reflects the Upper Body Cavity.
- The middle third of the tongue reflects the Middle Body Cavity.
- The root of the tongue reflects the Lower Body Cavity.

12.1 The Tongue Fur

Table 12.1: The Tongue Fur

Appearance	Meaning
Thick	Chronic difficult condition or latent EPA.
Short	deficiency essence or *qi*. SP.6 & KID.3 reinforce.
No Fur (bald)	Depletion of essence. Tian Men Dong, KID.3 Reinforce.
White	Cold/poor function. Bai Zhu (MBC) and Du Zhong (LBC).
Dry White caked	Stag. of phlegm/food & deficiency of fluid. ST.36 reduce, KID.3 reinforce.
White	Shows heat in pulse but cause patient is weak they cannot show yellow fur.
Slight Yellow	Slight heat or strong heat but weak health. TW.5 reduce.
Deep Yellow	Strong heat. Huang Qin TW.5 & GB.41 reduce.
Slimy porridge	Food/phlegm stagnation damp. Zhi Ke, Zhi Shi, ST.36, CV.12 reduce.
Dry	Heat drying off fluids/constipation - Sang Shen TW.5 reduce KID.3 reinforce.
Moist	Damp. Bai Zhu & Fu Ling SP.6 reinforce.
Moist	Def/*yin* heat. Xuan Shen, Nu Zhen Zhi, TW.5 reduce KID.3 reinforce.

12.2 The Tongue Tastebuds

(Often relating to the glands)

Table 12.2: The Tongue Tastebuds

Appearance	Meaning
Red	Heat. Qing Dai, TW.5 reduce.
Raised red	Inflammation. Zhi Zi, LIV.3 reduce.
Sunken	Depleted. Xuan Shen, Nu Zhen Zi, Dang Shen (Codonopsis), SP.6 & KID.3 reinforce.
Pale	deficiency. Nu Zhen Zhi, He Shou Wu, Dang Gui, Bai Zhu SP.6 & KID.3.
Sunken Pale	Depleted & Deficient. Nu Zhen Zhi, He Shou Wu, Dang Shen (Codonopsis), Bai Zhu, Dang Gui. Reinforce SP.6, KID.3, KID.6, BL.23, BL.20, BL.13.

12.3 The Colour of Tongue Substance

Table 12.3: The Colour of Tongue Substance

Appearance	Meaning
Light Pink/Pale	Deficiency of Blood/*qi*/essence. Nu Zhen Zhi, He Shou Wu, Bai Zhu, Dang Gui.
Ashy	Weak/cold/viral/damp/EPA toxin cause def blood stagnation. Dan Shen (Miltiorrhizae).
Blue	Chronic blood impairment use Zao Xiu & Dan Shen (Miltiorrhizae).
Red	Heat. Huang Qin, TW.5 reduce.
Deep Red	Strong heat. Qing Dai, Mu Dan Pi, SP.10 reduce.
Red Purple	Excess heat cause blood stagn. Huang Qin, Mu Dan Pi, Tao Ren. GB.40 reduce.
Purple	Alcohol or drug intoxication. Zao Xiu, Da Huang, Gan Cao.
Dark Purple	Blood Clot. E Zhu, San Leng, Tao Ren.
Grey surface	Blood stag/EPA. Mu Dan Pi.
Black surface	Kid essence def. Tian Men Dong, Nu Zhen Zhi, Ji Nei Jin. SP.6 & KID.3 reinforce.

12.4 The Tongue Surface Condition

(reflects the health of the whole body)

Table 12.4: The Tongue Surface Condition

Appearance	Meaning
Dry	Heat drying off body fluids. Bai He, Shi Hu, Lu Gen, TW 5 reduce K3 reinforce.
Moist	Damp. Bai Zhu & Fu Ling , SP.6 reinforce.
Moist	Deficiency *yin* heat. Xuan Shen & Nu Zhen Zi. SP.6 and KID.3 reinforce.
Local Fissure	Chronic impairment.
Beef Steak	Serious whole body impairment due to long term drug intox/hereditary factors.

12.5 The Tongue Shape

Table 12.5: The Tongue Shape

Appearance	Meaning
Thin	Def body constitution.
Tooth marks	Sp. *qi* def. absorption problem. If pale: Bai Zhu. If red: Liver heat - Zhi Zi.
Mouth Ulcer	Toxins and latent heat in Stomach. Pu Gong Ying, Da Huang, ST.36 reduce.

Table 12.6: Tongue Diagnosis

Fur	Substance	Character	Area	Meaning	Herbs
white			root	cold/poor function LBC	Du Zhong, Dang Shen (Codonopsis).
	red			blood heat	Mu Dan Pi, Chi Shao Yao.
	pale	fissure	root	blood and def kidney	Shu Di Huang, Fu Pen Zi.
	red	fissure	root	blood heat kidney	Nu Zhen Zi, Chi Shao Yao.
		fissure	tip	Heart stress damaged	Fu Shen, He Huan Pi.
yellow				heat	Huang Qin.
yellow		yogurt	mid	heat food stag	Shen Qu, Zhu Ru, Shan Zha
yellow		dry	mid	SP, ST, LIV MBC Heat dries Bwls	Huang Qin, Bai He, Sang Shen
yellow		dry	root	heat in LBC	Nu Zhen Zi, Chi Shao Yao
slimy			root	phlegm/food stag	Zhi Ke, Qing Pi.
no fur	pale		tip	def lung fluid	Dang Shen (Codonopsis), Lu Gen, Ye Jiao Teng
no fur	red		tip	dried lung essence	Shi Hu, Bai Ji, Bai He, Lu Gen.
no fur	pale		root	kid/blood def	Nu Zhen Zi, He Shou Wu.
no fur	red		root	heat impairing essence	Nu Zhen Zi, Chi Shao Yao.
no fur	red	big buds	tip	Heart heat	Xie Bai, Gua Lou, Zhu Ru, Bai He
	red	big buds	mid	liv heat MBC heat	Zhi Zi, Shen Qu, Fu Ling
	pale	big buds	root	def blood hormone	Dang Gui, He Shou Wu
	red	big buds	root	hormone GU heat	Chi Shao Yao, Mu Dan Pi
	pale	big buds	tip	def impaired glands	Xuan Shen, Nu Zhen Zi
black				dehydration	Tian Men Dong, Sang Shen.
	black		mid	organ blood clot	Tao Ren, Dan Shen (Miltiorrhizae).
	purple			blood toxicity	Zao Xiu, Ban Zhi Lian.
	pale	sml buds	tip	weak immunity	Dang Shen (Codonopsis), Shu Di Huang.
	red	big buds	tip	blood/Heart heat	Xie Bai, Gua Lou, Xuan Shen, Bai He
	red		sides	Liver heat	Zhi Zi
		prolapsed	sides	weak SP poor function	Bai Zhu, Fu Ling, Gan Cau
	dark veins		under	toxins in bowels	Da Huang, if Blood stagn Tao Ren

Chapter 13

Lip Diagnosis

(Spleen Blood/Vessels)

- **Lips pale pink**: Weak blood & insufficiency of red blood cells and essence with poor function of Spleen, use Bai Zhu, Ji Xue Teng, He Shou Wu, Shu Di Huang and Dang Gui. Could balance with Mu Dan Pi & Bai Shao Yao also check & see if Kidneys need tonifying. Reinforce SP.6, KID.3, KID.6, BL.23, BL.20, BL.13.
- **Both lips dark red & dry**: Blood heat & dryness use Mu Dan Pi, Bai Shao Yao, Tian Men Dong, Bai He, Gan Cao to balance.
- **Tongue pale pink and lips deep red**: Acute heat impairing blood composition causing anaemia. Use Mu Dan Pi, Sheng Di Huang, Shu Di Huang, Dang Gui.
- **Lips are dark brown (pigs Liver)**: Heat has impaired Spleen tissue & function, essence and blood is damaged. Use Dan Shen (Miltiorrhizae), Tao Ren, Hong Hua, Dang Gui.
- **Outside edges of lips are cracked**: Prolonged heat damaging Spleen tissue/function, use Mu Dan Pi, Tian Men Dong and Gan Cao.
- **Cracks have burst with blood stains**: Intense blood heat leading to bleeding. Use Sheng Di Huang, Mu Dan Pi and Dan Shen (Miltiorrhizae), Bai Ji and Bai Shao Yao.
- **Upper half of lower lip ashy**: Blood stasis Tao Ren, Chi Shao Yao and Dan Shen (Miltiorrhizae).
- **Outside of lips go ashy**: Blood stasis in Spleen, Liver and blood vessels. Use Tao Ren, Dan Shen (Miltiorrhizae), Hong Hua, and E Zhu.
- **Lips are purple**: Indicates a toxin in the blood, use Pu Gong Ying or Ban Zhi Lian or Zao Xiu & add Da Huang to purge toxins from system.

Around Lips

- **Blue tinge**: Wind trapped in digestive intestinal system causing disruption of digestion, malabsorption, bloating & flatulence. Use Hai Feng Teng or Fang Feng and Zhi Ke.
- **Green tinge**: Indigestion use Shen Qu, Zhi Shi and Yu Jin.
- **Yellow tinge**: Malabsorbtion leading to malnutrition of blood use Bai Zhu and Ji Xue Teng.

Chapter 14

Eye Diagnosis

Overall the eye reflects the Liver. Yet the sclera pertain to the Lungs, iris to the Liver, and pupil to the Kidneys.

14.1 Cornea

The cornea is the lens that covers the whole of the eye. If it is very noticeably glossy this indicates mid or old age deficiency or loss of essence or lack of sleep or kidney exhaustion so we use Nu Zhen Zi.

14.2 Pupil

On pupil reflex test, if:

- **Pupils constrict**: weak Kidneys. Use Fu Pen Zi and or Du Zhong
- **Pupils dilate**: nervous tension. Use He Huan Pi
- **Pupils unresponsive**: nerves dulled by tranquilizers, drugs or alcohol. Use Zao Xiu or Ban Zhi Lian, Da Huang and reduce intoxicant
- **One pupil dilates & other constricts**: One eye weaker than the other is a prescription problem

14.3 Sclera

- **Red**: Heat, Use Mu Dan Pi. EPA or whole body heat, Use Huang Qin. Liver / eye heat, use Ju Hua
- **Blue**: Wind, Use Hai Feng Teng or Fang Feng
- **Green**: Indigestion, Use Shen Qu and Ji Nei Jin
- **Cloudy**: long term Liver damage from spicy food or due to old age use Huang *qi*, Nu Zhen Zi, Chai Hu and Tao Ren, Yin Chen Hao
- **Yellow**: Liver problem, Use Huang *qi* if deficient, or Zhi Zi if excessive
- **Dark yellow**: jaundice, Use Huang *qi* and Zhi Zi, Yin Chen Hao

14.4 Iris

- **Transparent and clear** : Liver in good condition
- **Dark and uneven**: Liver problem
- **White ring**: Liver impairment or old age regression

14.5 Eye Lids

- **Red under lower eye lid**: heat in blood. Use Mu Dan Pi & Shan Zhu Yu to balance.
- **Pale under lower eye lid**: blood and red blood cell deficiency (anaemia), Use Dang Gui, Shu Di Huang, Bai Zhu, He Shou Wu, Ji Xue Teng.
- **Yellow under lower eye lid**: bile in blood relating to Liver function. Use Huang *qi*.
- **Conjunctivitis**: Use Xia Ku Cao, Huang Lian, Lian Qiao, Sang Ye, Ju Hua, Bo He, Chai Hu, Gan Cao.

Chapter 15

Diagnosis by Excretions and Secretions

15.1 Mouth saliva

- **Excess clear thin** : weak Spleen or excess damp. Use Bai Zhu, Fu Ling.

15.2 Nasal discharge

- **Excess thin clear** : weak Spleen; use Bai Zhu. Excess damp / EPA onset symptom; use Bo He, Bai Zhu.
- **Yellow** : heat EPA. Use Huang Qin, Lu Gen.
- **Thick yellow** : heat EPA late stage. Use Xuan Shen, Zhu Ru, Xin Yi and Tao Ren.
- **Green** : bacterial agent present. Use Jin Yin Hua and Da Huang.
- **Hard brown** : chronic latent heat. Use Tian Men Dong and Lu Gen.
- **Red (blood)** : heat–inflammation and dryness cause fracture of blood vessels in Lungs. Use Bai Ji, Bai He, Xuan Shen, Shi Hu, Sheng Di Huang.

15.3 Eye discharge

- **White** : wind & cold. Use Jing Jie.
- **Yellow** : wind and heat. Use Ju Hua.

15.4 Anus/Stools

- **Pale & soft faeces** : weak Spleen or excess damp. Use Bai Zhu, Fu Ling, Gan Cao, and Shan Yao.
- **Undigested food in faesces** : weak Spleen, weak digestion absorbtion system. Use Bai Zhu, Fu Ling, Ji Nei Jin, and Shen Qu.

- **Diarrhoea** : weak Spleen/digestive absorption. Use Bai Zhu, Gan Cao, Shan Yao. For food poisoning, use Pu Gong Ying, and Gan Cao.
- **Hard and dark faeces** : heat and dryness. Use Sang Shen, Tao Ren, and Xing Ren. For elderly use Dang Gui.
- **Constipation** : chronic heat and dryness. Use Huang Qin, Sang Shen, Tao Ren, and Xing Ren.
- **Blood in faeces** : heat, inflamation and dryness cause fracture of blood vessels. Use Di Yu Tan, Zhi Zi Tan, Sheng Di Huang, Bai Mao Gen, Pu Huang, and Huai Hua.
- **Blood** : heat, inflamation and dryness cause fracture of blood vessels. Use Sheng Di Huang, Chi Shao Yao, Shan Zhu Yu, Nu Zhen Zi, Mu Dan Pi, Di Yu Tan, Zhi Zi Tan, Bai Mao Gen, Pu Huang and Huai Hua.
- **Black moist stool** : Stomach or duodenal ulcer causing internal bleeding. Use Gan Cao, Bai Ji, Hai Piao Xiao, Mo Yao and Ru Xiang

Bleeding from anus and black stool could also be bowel cancer.

15.5 Urine

- **Excessive and pale urine** : weak Bladder and Kidneys. Use Fu Pen Zi, Jin Ying Zi.
- **Sperm in urine** : very weak Bladder, Kidneys. Use Fu Pen Zi, Jin Ying Zi, Shan Yao.
- **Scanty, Dark yellow urine** : Heat/dryness in Bladder. Use Bai Mao Gen, Lu Gen, Che Qian Cao, Chi Shao Yao.
- **Scanty, brown urine** : LBC Chronic heat & dryness. Use Bai Mao Gen, Lu Gen, Hua Shi, Tian Men Dong, Che Qian Cao.
- **Red blood in urine** : heat–Inflamation and dryness causing fracture of blood vessels (possible Bladder cancer get checked). Use Di Yu Tan, Zhi Zi Tan, Lu Gen, Bai Mao Gen and Nu Zhen Zi.
- **Intermittent urination in men** : due to enlarged prostrate. Use E Zhu and San Leng, Che Qian Cao, Pu Gong Ying, Chi Shao Yao, Mu Li, Lu Gen, Zao Jiao Ci and general health. Also recommend Saw Palmetto and Zinc supplements.

15.6 Vaginal Discharge and Menses

- **Whiteish discharge (leucorrhea)** : damp weak Spleen and immune system. Use Qian Shi, Yi Yi Ren, Jin Ying Zi, Fu Pen Zi, Shan Yao, Bai Zhu, Fu Ling.
- **Yellow discharge** : heat and infection. Use Pu Gong Ying, Jin Yin Hua, Bai Zhi, Nu Zhen Zhi, Huai Hua, Chi Shao Yao, Gan Cao.
- **Scanty, thin pale menstrual flow** : Weak blood. Use Ji Xue Teng, He Shou Wu, Bai Zhu, Dang Gui, Chuan Xiong. Prescribe after menstruation.
- **Darker, thicker flow** : blood heat. Use Chi Shao Yao, Sheng Di Huang, Mu Dan Pi and Bai Shao. Prescribe before menstruation.

- **Coffee coloured clots** : blood thickening. Use E Zhu, San Leng, Tao Ren and Dan Shen (Miltiorrhizae).
- **No menstruation (amenorrhea)** : blood stagnation. Use San Leng, E Zhu, Chai Hu, Tao Ren. Prescribe before menstruation.
- **Excessive bleeding (hypermenorrhea) with dark blood** : blood heat in lower body. Use E Jiao, Di Yu Tan, Zhi Zi Tan, Sheng Di Huang, Bai Ji. Prescribe after menstruation.

Chapter 16

Pulse Diagnosis

When taking the pulse the practitioner places the index, middle, and ring fingers on the radial artery. Three degrees of pressure, light touch, medium touch and heavy touch are applied this corresponds to UBC, MBC, LBC and the pulse is then described as having been found on the Floating, Moderate or Deep level.

16.1 Pulse Positions

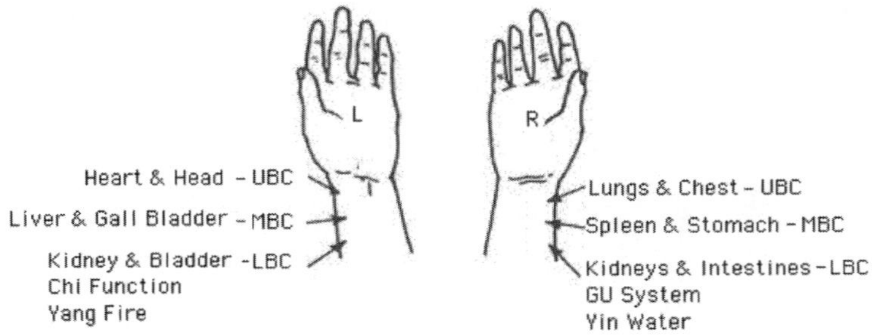

Each of the pulses is a place where we can take a reading about what is happening to a section of the body and the corresponding internal organs.

Table 16.1: Pulse Positions

	Pulses - Left Wrist	**Pulses - Right Wrist**
Stage One (UBC)	Heart & Head	Lungs & Chest
Stage Two (MBC)	Liver & Gall Bladder	Spleen & Stomach
Stage Three (LBC)	Kidney & Bladder Chi Function *yang* Fire	Kidneys & GV.4 Intestines & Genital Urinary *yin* Water & Triple Warmer Lymphatic system Glandular system

The pulses can be categorised into five groups;

Table 16.2: Pulse Categories

Pulse Level	Pulse Speed	Pulse Volume	Pulse Resistance	Pulse Condition
Superfical	Rapid	Full	Strong	Tense
Floating	Fast	Thin	Weak	Taut/Wiry
Moderate	Slow	Hollow	Soft	Hesitant
Deep		Thready	Firm	Tremulous
Hidden				

Pulse Level

Is the pulse floating, moderate or deep? If the pulse is floating on the Lung pulse this could be EPA onset. The pulse can be floating in old people due to old age exhaustion, usually hollow - no energy or resistance. A moderate level can signify a 'normal' pulse. A deep pulse could mean a chronic condition or indicate interior syndromes. If the pulse is deep and forceful, it indicates interior syndromes with excessive type. If the pulse is deep and weak, it indicates interior syndromes of a deficiency type.

Pulse Speed

Is the pulse rapid, fast or slow? If the pulse is rapid, fast and urgent this could mean an overactive metabolism with heat in the whole system. deficiency *yang* heat presents a rapid and weak pulse. If the pulse is slow this could point to cold/poor function and an underactive metabolism. *qi* contracts and blood flow stagnates on exposure to cold.

Pulse Volume

Is the pulse hollow ? Like a spring onion, when you squeeze, it feels empty inside. Hollow pulse means deficiency of blood and essence. Is the pulse full? Heat raises the volume of the blood so it would feel full. Is the volume fine and thready? This indicates deficiency of *qi* and blood.

Pulse Resistance

The pulse resistance gives us an indication of the strength of the *qi*. When you press the pulse, if it resists the pressure this strong pulse resistance is an indicator of strong pulse energy. When you lift your finger up after having applied pressure if the pulse follows your finger up this is also an indicator of strong *qi*. However if when you press the pulse it collapses under the pressure and shows no resistance and fails to follow your finger up when you reduce the pressure on the pulse then this indicates weak pulse resistance and so weak *qi*. If the resistance is firm, a solid feeling this could indicate an abnormal hardening of the tissues.

Pulse Condition (Character)

Here we can get an exact indication with what is happening in specific areas of the body depending on which stage the condition/character is. For example a hesitant pulse on stage one left would indicate blood damaged by long term heat in the Heart. Slippery on stage one right would be stagnation of phlegm in the Lungs.

16.2 Description of Pulses

- Slippery : feels like a rolling pearl in a basin, very fluid and full.
- Choppy : has no strength and is irregular.
- Full : large and rounded, can be felt at all levels.
- Empty : hard to detect, felt only slightly on superficial level when pressure applied.
- Slow : slower than the normal rate of four to five beats per breath.
- Rapid : six to seven beats per breath.
- Superficial : easily felt on the skin surface.
- Deep : only felt with a heavy touch.
- Fine : thread like thin.
- Hollow : like a spring onion stalk.
- Long : long beat covers stage 1, 2 and 3, feeling long and firm.
- Soft: thread on water, no resistance.
- Tense : like a vibrating string.
- Taut/Wiry : actual artery pushing out, expanding, abnormal circulation in energy.

16.2.1 Some Samples of Pulses and their Meaning

Floating

A pulse becomes floating from both an exterior syndrome, or a deficiency syndrome. In the case of an exterior syndrome, the *wei qi* (defensive *qi*) rises to the surface to defend the body from the pathogen. It brings blood and qi to the surface, therefore the pulse is felt mostly at the superficial level. If a deficiency syndrome exists, the *yin* is unable to anchor the *yang*, or the nearby *yang* is so weak it nears exhaustion. In both cases the *yang* will float to the body's surface as the separation of *yang* and *yin* begins.

Deep

This pulse occurs either through interior excess or deficiency. In the case of interior excess; Pathogenic factors accumulating in the body, they will obstruct the outward movement of *qi*, blood and *yang*. If through interior deficiency, *qi* and blood are insufficient and therefore unable to fill the channels.

Rapid

A rapid pulse can be caused by excess pathogenic heat increasing the circulation of *qi* and blood. *yin* and blood deficiency lead to a relative *yang* excess and will also cause the pulse to quicken.

Forceful

An excessive pulse can be from the hyperactivity of pathogenic factors and strong anti-pathogenic *qi*. As pathogenic and anti-pathogenic factors struggle, qi and blood accumulate filling the vessels.

Feeble

A feeble pulse is a sign of *qi* and blood deficiency.

Tense

This pulse can be caused by *qi* and blood stagnation, phlegm, pain and a retained pathogen.

16.3 How to take the Pulse

The patient has been lying or sitting down for at least five minutes so that they can be calm and so that the pulse can give us a clear reading.

We ask the patient about their symptoms and observe the eyes, lips, face and tongue. We may have formed an opinion of the patients condition but before we decide on a treatment strategy the pulse should be read to verify the diagnosis.

We feel the pulse on the radial artery of the wrists by placing three fingers side by side with the middle finger proximal to the styloid process of the wrist.

We do a general pulse diagnosis on the right side.

We then do a general pulse diagnosis on the left side.

We then do a detailed pulse diagnosis on the right side.

We then do a detailed pulse diagnosis on the left side.

After this initial overall pulse reading we should have a picture of the general state of the body, whether the condition is one of excess or deficiency and if there is an external pathogen affecting the body. If deficient the pulse will be weak, fine, feeble or hollow, if in excess it will be forceful, full or rapid.

We then compare the information from the pulse diagnosis with the information from the tongue and eyes, lips and face as well as the patients description of their symptoms.

If the pulse matches the tongue and eyes, lips and face as well as the patients description of their symptoms then we can formulate our acupuncture and herbal treatments. If it does not match then it means we are dealing with a very complicated condition and need to spend more time diagnosing to be clear how to proceed.

Pulse diagnosis is important as it gives us current information about the state of the patients blood, essence and *qi* and which internal organs have been affected by wind, heat, damp, dryness, cold and phlegm.

16.3.1 The Pulse and the Seasons

In spring, the Liver pulse may be affected by the wind.
In summer, the Heart pulse may be affected by the heat.
In the late summer, the Spleen pulse may be affected by the damp.
In the autumn, the lung pulse may be affected by the dryness.
In the winter, the kidney pulse may be affected by the cold.

16.3.2 Pulse level

Stage One pulses should be found on the superficial level
Stage Two pulses should be found on the moderate level
Stage Three pulses should be found on the deep level

If any pulse is found on a level that does not reflect its position in the body it indicates an imbalance in the bodies health.

Superficial pulse

A superficial pulse may indicate the invasion of an external pathogen. In this case the pulse is superficial because the bodies *qi* is rising up to fight the EPA.

A superficial pulse could also indicate a deficiency condition, it has lost its root and is floating up without any anchor to hold it.

Deep pulse

A deep pulse indicates that either the problem is now located deeper in the body, usually found in chronic conditions, or the *qi* is going to where the problem is located so we find the pulse on the deep level.

A deep pulse could also due to the collapse of the bodies strength and so it does not have the power to rise any higher, this is a type of deficiency pulse.

16.3.3 deficiency Pulses

In deficiency states a hollow pulse (like an empty hose) means depletion of essence and blood or if combined with taught it means exhaustion of sperm. A fine pulse (threadlike) is a lack of blood and a feeble pulse (faint) is exhaustion with lack of blood and essence. A small pulse (not rising to the fingers) also indicates weak general health and deficiency. In deficiency conditions if the pulse is slow it means weakness and cold/poor function. If soft (thread on water) with no resistance this is damp, water retention.

16.3.4 Excess Pulses

In excess conditions a rapid pulse (urgent) represents EPA, heat, temperature or inflammation, a forceful pulse (pushing forwards, like wave dashing on the shore) shows heat stagnation (latent heat), a full pulse means blood heat or high blood pressure and a long pulse may also indicate heat. If the pulse is taut/wiry (like a spring) particularly in the middle this represents Liver problems or pain, if hesitant (chopstick over bamboo) this indicates blood heat or inflammation, if intermittent this is a Heart problem and if viscous (sluggish and oily) this indicates blood thickening due to chronic heat or the menopause.

16.3.5 Acute and latent EPA Pulses

If the pulse is more forceful on the right than the left then the EPA is acute where as if stronger on the left this indicates a chronic/latent EPA. This idea can be remembered with the expression, Right Recent, Left Latent.

A tense pulse (two discordant strings) is often felt when there is an EPA, it is the internal organs being aggravated by the pathogen.

Another EPA pulse that may be combined with the tense pulse is the wavy pulse also sometimes called a divergent pulse, the feeling is like a wave as the pulses land one after the other rather all at the same time like a normal pulse. A wavy pulse can also represent a hormonal imbalance or stress.

16.4 Pulse Combinations

Pulses are found in combinations:

Rapid Pulse – heat

Rapid + tension = heat and EPA
Rapid + short = heat and weak *qi*
Rapid + intermittent = heat and Heart valve problems
Rapid + soft = heat and water retention
Rapid + with pulse energy = excessive heat with a strong constitution
Rapid + without pulse energy = excessive heat with a weak constitution

Slow Pulse – deficient organ / *yang qi* trapped internally

Slow + with pulse energy = abdominal pain due to cold stagnation
Slow + without pulse energy = deficient cold disease due to insufficient *yang qi*

Feeble Pulse – deficiency

Feeble + hollow = blood /essence/ *qi* exhaustion
Feeble + short/small = *qi* deficiency

Feeble + wavy = period blood loss
Feeble + viscous = Heart blood impairment

Floating pulse – external disease / internal deficiency

Floating + pulse energy = external wind attack
Floating + without pulse energy = internal blood def
Floating + slow = *qi* deficient person injured by wind
Floating + rapid = external wind and heat
Floating + tense (tight) = wind and cold stagnating in channels
Floating + moderate = damp stagnated in muscles
Floating + empty = summer heat injury to kidney essence
Floating + hollow = great loss of blood
Floating + forcefull = *yin* deficient excess fire
Floating + fine = fatigue
Floating + soft = deficiency or essence damage
Floating + scattered = deficiency of blood and *qi*
Floating + taut/wiry = excessive internal phlegm
Floating + slippery = excessive stagnation of heat phlegm

Deep Pulse – chronic organ disorders

Deep + pulse energy = phlegm affecting digestion
Deep + without pulse energy = *qi* stagnation
Deep + slow = deficient cold
Deep + rapid = heat trapped internally
Deep + tense = stagnant cold causing pain
Deep + moderate = accumulated thin phlegm
Deep + firm = chronic cold disease
Deep + full = excessive internal heat
Deep + weak = essence damaged
Deep + thin = damp stagnation
Deep + taut/wiry = pain caused by phlegm
Deep + slippery = food phlegm stagnation and weak digestion
Deep + hidden = vomiting and diarrhoea due to damp toxin

16.5 Herbs for Pulses

Pulse	Pathology and Herbs Used
Hollow-feeble	Exhaustion blood, essence or energy. Use Nu Zhen Zi, Dang Shen (Codonopsis), He Shou Wu.
Rapid-tense	Exogenous-temperature. Use Huang Qin, Da Huang, Shen Qu, Jing Jie.
Rapid-short	Temperature-weak vital energy. Use Qing Dai, Shan Yao.
Slow-intermittent	Heart valve problem with cold function. Use Xie Bai, Gua Lou, Dang Shen (Codonopsis).
Rapid-intermittent	Heart valve problem with heat. Use Xie Bai, Gua Lou, Zhu Ru.
Rapid-soft	Heat with water retention. Use Huang Qin, Bai Zhu, Fu Ling.
Slow-soft	Cold function with water retention. Use Du Zhong, Bai Zhu, Fu Ling.
Long-taut	Liver heat. Use Chai Hu, Yu Jin, Zhi Zi.
Wavy-tense	Exogenous toxin. Use Gui Zhi, Mu Dan Pi, Xuan Shen, Da Huang.
Wavy-taut	Pain. Use Yu Jin, Yan Hu Suo.
Full-forceful	Heat-food stagnation. Use Huang Qin, Da Huang, Shen Qu.
Feeble-short	Deficient-exausted vital energy. Use Shan Yao, Shu Di Huang, Dang Shen (Codonopsis).
Feeble-wavy	Menstrual blood loss. Use Sheng Di Huang, Shu Di Huang, Nu Zhen Zi, He Shou Wu.
Viscous-slow	Thickened blood-poor circulation. Use Tao Ren, Shui Zhi, Chi Shao Yao, E Zhu, San Leng.
Hollow-taut	Exaustion of sperm. Use Du Zhong, Tian Men Dong, Shang Shen.
Fine-forceful	Heat-food stagnation with blood def. Use Shu Di Huang, Sheng Di Huang, Shen Qu, Huang Qin.
Tense-hesitant	Spontaneous bleeding. Use Sheng Di Huang, Di Yu, Bai Ji, Lu Gen, Xuan Shen, Mu Dan Pi.
Feeble-viscous	Heart-blood impaired post Heart attack. Use Dang Shen (Codonopsis), Tao Ren, Shu Di Huang, Sheng Di Huang.
Taut-tense	Toxin producing distention. Use Da Huang, Mu Xiang, Yan Hu Suo.
Forceful-intermittent	Heart-valve and kidney impaired. Use Xie Bai, Gua Lou, Xian Mao.

16.6 Pulse Applications

Hollow

- **Description** ; spring onion stalk
- **Feeling** ; empty in center
- **Meaning** ; depletion of essence, blood or sperm
- **Herbs** ; Nu Zhen Zi, Shu Di Huang, Sheng Di Huang.
- **Points** ; Reinforce - SP.6, SP.9, KID.3, BL.23, BL.20, LIV.8.

Fine

- **Description** ; thread like thin
- **Meaning** ; lack of blood
- **Herbs** ; Dang Gui, Sheng Di Huang, Shu Di Huang
- **Points** ; Reinforce BL.15, BL.20, SP.6, LIV.8.

Feeble

- **Description** ; thready / weak / faint
- **Feeling** ; lack of blood, essence or energy
- **Meaning** ; exhaustion
- **Herbs** ; Dang Shen (Codonopsis)
- **Points** ; Reinforce CV.4, BL.20, BL.23

Rapid

- **Description** ; fast
- **Feeling** ; urgent
- **Meaning** ; temperature-heat-virus
- **Herbs** ; Huang Qin
- **Points** ; Reduce TW.5, GB.41 - Trigram points to clear heat.

Forceful

- **Description** ; accent forward thrust, like waves pounding shore
- **Feeling** ; pushy
- **Meaning** ; heat-stag. food / mass-hot flush-constipation
- **Herbs** ; Huang Qin, Da Huang
- **Points** ; Reduce TW.5, ST.36, LI.4.

Full

- **Description** ; artery full of blood
- **Feeling** ; thick / firm
- **Meaning** ; excess heat-Heart pumping hard-HBP with high grade temperature
- **Herbs** ; Huang Qin, Di Yu
- **Points** ; Reduce TW.5, GB.41, apex of ear (SB), LI.1 (SB), LU.11 (SB).

Long

- **Description** ; long beat covers stages 1, 2 and 3
- **Feeling** ; long / firm
- **Meaning** ; strong health or heat, healthy or flushed
- **Herbs** ; nothing or Huang Qin
- **Points** ; Reduce TW.5, GB.41 - Trigram points to clear heat.

Short

- **Description** ; short beat unable to cover Stage One.
- **Feeling** ; short/fine
- **Meaning** ; weak health - deficiency asthenia exhaustion - anaemic
- **Herbs** ; Shan Yao, Shu Di Huang, Sheng Di Huang
- **Points** ; Reinforce CV.4, SP.6, SP.9, SP.10, KID.3, ST.36.

Taut/Wiry

- **Description** ; string tight
- **Feeling** ; artery tight/expanding
- **Meaning** ; pain - Liver disorder/disease - distention - spasm - IBS - poor quality blood.
- **Herbs** ; Yu Jin, Yan Hu Suo
- **Points** ; Reduce LIV.3, ST.36, GB.34.

Hesitant

- **Description** ; scraping, like a chopstick along bamboo
- **Feeling** ; wash board-many fine/short beats
- **Meaning** ; blood heat - inflammation blood stasis - anaemic glands
- **Herbs** ; Mu Dan Pi, Chi Shao Yao
- **Points** ; Reduce SP.10, LIV.2, GB.40.

Tense

- **Description** ; 2 discordant strings vibrating
- **Feeling** ; tense wave on that stage
- **Meaning** ; exogenous toxin aggravating organs - viral/bacterial invasion
- **Herbs** ; Shen Qu, Da Huang
- **Points** ; Reduce LU.7, TW.5, LI.4. Reinforce KID.6.

Slow

- **Description** ; retarded
- **Feeling** ; slower
- **Meaning** ; weakness-cold function-weakness-asthenia-athlete-weak Heart
- **Herbs** ; Shan Yao, Ye Jiao Teng, Dang Shen (Codonopsis), Du Zhong
- **Points** ; Reinforce BL.15, BL.20, BL.23.

Intermittant – Regular/Irregular

- **Description** ; 3 beats miss one then quick double
- **Feeling** ; thicken Heart valvular problem-enlargment of Heart
- **Meaning** ; mitral valve stenosis-hypertrophy
- **Herbs** ; Xie Bia, Gua Lou
- **Points** ; Reinforce B.15 and BL.23.

Soft

- **Description** ; thread on water
- **Feeling** ; no resistance
- **Meaning** ; damp-water retained in blood/muscle-rheumatism-oedema
- **Herbs** ; Bai Zhu, Fu Ling, Fu Ling Pi
- **Points** ; Reinforce SP.6

Viscous

- **Description** ; sluggish / oily
- **Feeling** ; slurs / blood feels thick
- **Meaning** ; blood thickening-menopause-stroke prone
- **Herbs** ; Tao Ren, Shui Zhi
- **Points** ; Reduce SP.6, 9, 10, TW.5, LIV.5, GB.40 and GB.41.

Wavy or Divergent

- **Description** ; stages 1,2 and 3 are like a wave
- **Feeling** ; makes your 3 fingers rock
- **Meaning** ; hormones out of balance, or presence of toxins;
  - Menstruation
  - Ovulation
  - Pregnancy
  - PMS/PMT
  - Stress/man or woman
  - HRT patches/pills
  - Menopause
  - Toxic pathogens
  - Virus
  - Bacteria
  - Chemical pollutant
  - Pharmaceutical or recreational drugs
  - Alcohol/Tobacco (cigarettes, cigars, rollies and pipes)
  - Nicorette gum/patch
- **Herbs** ; Da Huang, Chai Hu, He Huan Pi
- **Points** ; Reduce GB.40, GB.34, LIV.3.